DISS ETH NO. 18491

Inverse Symmetry Breaking
in Low-Dimensional Systems

A dissertation submitted to

ETH Zurich

for the degree of

Doctor of Sciences

presented by

NICULIN ANDRI SARATZ

Dipl. Phys. ETH Zurich
born on March 21, 1979
citizen of Pontresina, GR

accepted on the recommendation of

Prof. Dr. D. Pescia
Prof. Dr. V. Pokrovsky
Prof. Dr. C. H. Back
Prof. Dr. A. Vaterlaus

2009

Bibliografische Information der Deutschen Nationalbibliothek

Die Deutsche Nationalbibliothek verzeichnet diese Publikation in der
Deutschen Nationalbibliografie; detaillierte bibliografische Daten sind
im Internet über http://dnb.d-nb.de abrufbar.

©Copyright Logos Verlag Berlin GmbH 2010
Alle Rechte vorbehalten.

ISBN 978-3-8325-2403-6

Logos Verlag Berlin GmbH
Comeniushof, Gubener Str. 47,
10243 Berlin
Tel.: +49 (0)30 42 85 10 90
Fax: +49 (0)30 42 85 10 92
INTERNET: http://www.logos-verlag.de

Contents

Zusammenfassung

Atomar dünne Eisenfilme auf der Kupfer (001)-Oberfläche stellen ein Modellsystem für Musterbildung auf Grund von konkurrierenden Wechselwirkungen dar. Solche Filme sind charakterisiert durch eine starke senkrechte Anisotropie, welche die Magnetisierung dazu zwingt, entweder senkrecht in die Filmebene hinein oder daraus heraus zu zeigen. Die kurzreichweitige, anziehende, Austauschwechselwirkung versucht benachbarte Spins parallel auszurichten. Diese Wechselwirkung wird durch die schwächere, aber langreichweitige, abstossende, Dipolwechselwirkung frustriert, die eine antiparallele Ausrichtung der Spins bevorzugt. Als Folge davon bricht die Magnetisierung in ein Muster aus mesoskopischen Domänen auf, in denen die Magnetisierung jeweils in die Ebene oder aus der Ebene zeigt. Ein externes Magnetfeld versucht die Magnetisierung parallel zu sich auszurichten und bevorzugt daher eine Sorte von Domänen.

In dieser Dissertation untersuchen wir die Domänenmuster in einem zweidimensionalen Parameterraum, der durch Temperatur und Magnetfeld aufgespannt wird. Die lokale Magnetisierung der Eisenschicht wird mit hoher Ortsauflösung in einem Rasterelektronenmikroskop mit Polarisationsanalyse der Sekundärelektronen gemessen. Messungen können bei variierender Temperatur T und in variablem Magnetfeld H durchgeführt werden. Die genaue Kontrolle über das angelegte Feld auf einem Niveau weit unter dem Erdmagnetfeld hat sich als äusserst wichtig für die Experimente herausgestellt. Die Domänengrösse L nimmt mit steigender Temperatur stark ab, während das Magnetfeld den Flächenanteil $f_{\uparrow(\downarrow)}$ beeinflusst, der durch die beiden Domänenarten belegt wird. Unser System erlaubt es daher, die Domänengrösse und den Flächenanteil unabhängig voneinander einzustellen.

Es is wohlbekannt, dass Streifendomänen den Gleichgewichtszustand ohne Magnetfeld darstellen. In dieser Arbeit finden wir einen Übergang zu kreisförmigen Domänen, wenn das Magnetfeld bei gleichbleibender Temperatur erhöht wird. Dieser Übergang ist umkehrbar und findet in einem geeigneten Temperaturbereich im thermischen Gleichgewicht statt. Über einem gewissen kritischen Magnetfeld findet ein zweiter Übergang statt, der zu einem gleichförmigen, gesättigten Zustand führt. Trotz der bedeutenden experimentellen Anstrengungen war es bisher nicht gelungen, den Übergang zu den Kreisdomänen in echt zweidimensionalen magnetischen Systemen im thermischen Gleichgewicht zu beobachten. Hier konnten wir detaillierte Erkenntnis über den Übergangsprozess auf mikroskopischer Skala gewinnen. Wird in konstantem Magnetfeld die Temperatur erhöht, so beobachten wir die Phasenabfolge Uniform – Kreisdomänen – Streifen. Beim Übergang vom gleichförmigen Zustand zu den Kreisdomänen wird die Translationssymmetrie bei der Erwärmung gebrochen und beim zweiten Übergang zu den Streifen wird auch die Rotationssymmetrie gebrochen. Das untersuchte System zeigt also mehrfache, systematische, inverse Symmetriebrechung in Gegensatz zum gewöhnlichen Phasendiagramm, welches reguläre Symmetriebrechung voraussagt.

Skalenverhalten und Universalität werden auch in der Umgebung der inversen Symmetriebrechung beobachtet. Im Einklang mit theoretischen Voraussagen finden wir den Übergang von Streifen zu Kreisdomänen bei einem universellen Wert des Flächenanteils, der unabhängig von Temperatur und Feld ist. Einfache Skalengesetze, welche das kritische Feld H_C, die Domänengrösse ohne Magnetfeld L_0 und den Flächenanteil f miteinander in Verbindung bringen, führen zu einem Zusammenfallen aller Datenpunkte im H-T-f-Raum auf eine einzige Gerade. Darüber hinaus können das beobachtete Phasendiagramm und die inverse Symmetriebrechung mit Hilfe dieser Skalengesetze aus Grundzustandsrechnungen vorausgesagt werden.

Der hier beobachtete fundamentale Unterschied zwischen den Phasendiagrammen für dicke und echt zweidimensionale Filme unterstreicht die entscheidende Rolle der Dimensionalität. Die Reduktion des Systems von 3 auf 2 Dimensionen führt reguläre in inverse Symmetriebrechung über.

Es wurde vorgeschlagen, dass langreichweitige Wechselwirkungen zu selbstverursachter Glasbildung in frustrierten Systemen führen. Tatsächlich haben wir bei tiefen Temperaturen ungeordnete stationäre Zustände beobachtet, welche als ein Glas aus eingefrorenen Domänen gedeutet werden können. Zeitabhängige Untersuchungen wurden durchgeführt um zu bestimmen, unter welchen Bedingungen das System den Gleichgewichtszustand erreichen kann. Die Relaxationszeit wurde in Abhängigkeit der Temperatur und des Feldes gemessen. Die Resultate deuten auf ein nicht-Arrhenius Verhalten der Relaxationszeit hin, eine typische Eigenschaft von glasbildenden Systemen.

Abstract

Atomically thin iron films on the copper (001)-surface constitute a model system for
pattern formation due to competing interactions. They are characterized by a strong
perpendicular anisotropy forcing the magnetization to point either out of or into the
film plane. The short-ranged, attractive, exchange interaction of quantum mechanical
origin tries to align neighbouring spins parallel. This interaction is frustrated by the
weaker, but long-ranged, repulsive, dipolar interaction preferring anti-parallel alignment of the spins. As a result of this frustration the magnetization breaks up into a
pattern of mesoscopic domains of alternating up and down magnetization. An external
field applied perpendicular to the film plane tends to align the magnetization parallel
to it and therefore favours one type of domains.

In this thesis we investigate the domain patterns in the two parameter space spanned
by temperature and magnetic field. The local magnetization of the film is measured
with high spatial resolution in a scanning electron microscope with polarization analysis of the secondary electrons. Measurements are performed at varying temperature T
and in a variable applied magnetic field H. Precise control over the applied field on a
level well below the earth magnetic field proved to be crucial for our experiments. The
domain size L in this system decreases strongly with increasing temperature, while the
magnetic field changes the area fraction $f_{\uparrow(\downarrow)}$ occupied by up (or down) magnetized
domains. Our system allows therefore to tune the domain size and the area fraction
independently.

It is well established that stripe domains constitute the equilibrium state in zero
field. Here we find a transition from stripes to circular bubble domains as the magnetic
field is increased at constant temperature. This transition is found to be reversible
and occurs in thermal equilibrium in an appropriate range of temperatures. Above
some critical field the sample undergoes a second transition to a uniform, saturated
state. Despite the considerable experimental effort, in truly two dimensional magnetic
systems the equilibrium transition from stripes to bubbles has not been observed
before. Here we present detailed insight into the transition process on a microscopic
scale. Upon increasing the temperature in a constant applied field we observe the
phase sequence uniform-bubbles-stripes. At the transition from uniform to bubbles
the translational symmetry of the domain pattern is broken upon heating and at the
second transition from bubbles to stripes also the rotational symmetry is broken. The
system therefore shows multiple, systematic, inverse symmetry breaking, in striking
contrast to the standard, symmetry breaking, phase diagram predicted before.

Scaling and universality are also found in the regime of inverse symmetry breaking.
In agreement with theoretical predictions, the transition from stripes to bubbles is

found to occur at a universal value of the area fraction that is independent of temperature or field. Simple scaling laws relating the critical fields H_C, the domain size in zero field L_0 and the area fraction f produces a collapse of all data points in H-T-f-space onto a single straight line. Moreover, using these scaling and universality properties, the observed inverse symmetry breaking phase diagram can be predicted from ground state calculations.

The fundamental difference between the phase diagrams for thick and truly two-dimensional films underlines the crucial role of dimensionality. The reduction of the system from 3 to 2 dimensions changes the character of the phase diagram from regular to inverse symmetry breaking.

It has been suggested that long-range interactions lead to self-generated glassiness in frustrated systems. Indeed, at low temperatures long-lived disordered states are observed, being an indication that the system may be frozen in a domain glass. Investigations in the time domain were performed to address the question of the conditions under which the system can be considered equilibrated. Measurements of the relaxation times as a function of temperature and field point towards non-Arrhenius behaviour, as is typical in glassy systems.

1 Summary of Results

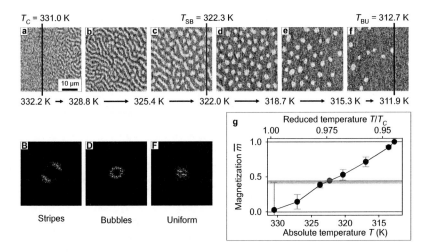

Figure 1.1: Observation of the bubble domain pattern and inverse symmetry breaking.

Figure 1.1 shows the sequence of magnetic domain patterns we observe in 1.9 atomic layers of Fe deposited at room temperature on a flat Cu(001) substrate. The sample is cooled in a constant applied field of 61 μT at a constant rate of -1 K/minute. Image a shows the transition from the uniform (U), contrastless, paramagnetic state to a stripe (S) pattern as the temperature falls below the transition temperature T_C, indicated by the black line. Although the stripe pattern in image b looks fairly disordered, it has a clear two-fold rotational symmetry and discrete translational symmetry. This becomes evident in its Fourier transform, image B, showing two peaks at $\pm\vec{q}_S$. In image c we observe a transition from stripes to bubbles (B) around a temperature T_{SB} (red line). The bubble phase is fully developed in image d. Its Fourier transform (D) shows a ring of finite radius $|\vec{q}_B|$, indicating that the rotational symmetry is restored upon cooling while the translational symmetry remains broken. Since some symmetry is restored as temperature decreases, the cross-over $S \to B$ marks a first Inverse Symmetry Breaking (ISB) transition. Additional cooling (images e, f) produces a second ISB transition from bubbles to the uniform state around a temperature T_{BU} (blue

1

line). The Fourier transform of image f, image F, shows a single peak centred at $\vec{q} = 0$, indicating that the full translational and rotational symmetries have been restored.

From each of the images a to f we can determine the area fraction $f_{\uparrow(\downarrow)}$ occupied by black (white) domains. The area fraction relates to a geometrical magnetization $\overline{m} = \frac{f_\uparrow - f_\downarrow}{f_\uparrow + f_\downarrow}$. In Fig. 1.1g the black dots shows the values of $\overline{m}$ computed for the images a to f. The red dot marks $T_{S \to B}$ at $\overline{m}_{S \to B} \approx 0.44$ and the blue dot marks $T_{B \to U}$ at $\overline{m}_{B \to U} \equiv 1$. We find that the value of $\overline{m}_{S \to B}$ is independent of temperature and the applied field.

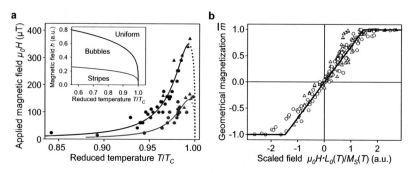

Figure 1.2: Pattern phase diagram and collapse plot.

Figure 1.2a shows the experimental pattern phase diagram in T-H-space obtained from measurements in constant field as discussed in Fig. 1.1 (circles) and from measurements at constant temperature (triangles). The transitions lines from stripes to bubbles (red) and from bubbles to uniform (blue) are fitted by a scaling law

$$H_t(T) \propto M_S(T)/L_0(T),\qquad(1.1)$$

where $L_0(T)$ is the equilibrium stripe width in zero field and $\pm M_S(T)$ is the constant magnetization inside the domains. We find that $H_{S \to B}/H_{B \to U} \approx 0.41$ is constant over the entire temperature range. Upward bending of the transition lines $H_t(T)$ is equivalent to ISB. The observed behaviour is in striking contrast to the standard phase diagram (inset) for micrometre thick films. The latter predicts down-bent transition lines and therefore regular Symmetry Breaking (SB) with the bubbles and the uniform state occurring at higher temperatures than the stripes.

Figure 1.2b shows the collapse of the data realized by a second scaling law

$$\overline{m}(T, H) \propto H \cdot L_0(T)/M_S(T).\qquad(1.2)$$

The individual measurements of $\overline{m}(T, H)$ obtained from series in constant field (circles) and at constant temperature (triangles) collapse onto a single straight line. The two scaling laws become equivalent if we consider the universality of $\overline{m}_{S \to B}$ and $\overline{m}_{B \to U}$.

Analytical and numerical ground state calculations explain the universal values of $\overline{m}_{\mathrm{SB}}$ (gray bar in Fig. 1.1g) and $H_{S \to B}/H_{B \to U}$ as well as the scaling laws (1.1) and (1.2). Inverse symmetry breaking results from (1.1) because over most of the temperature range $L_0(T)$ decreases strongly with increasing temperature while $M_S(T)$ varies only weakly (solid lines in Fig. 1.2a). Very close to T_C the domain size $L_0(T)$ becomes constant and $M_S(T)$ decreases strongly, leading to putative regular symmetry breaking (dotted lines). This part of the phase diagram is however not resolved in our experiments. For thick films the phase diagram is ruled by $H_t(T) \propto M_S(T)$ and no ISB is observed.

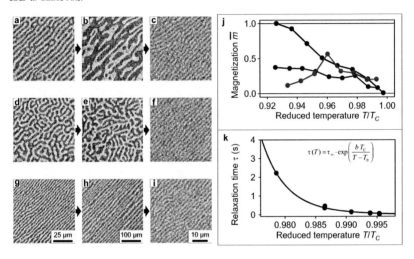

Figure 1.3: Non-equilibrium properties and relaxation times.

Self-generated glassiness has been predicted in frustrated sytems with weak long-range interactions, trapping the system in long-lived metastable states and impeding the relaxation to the equilibrium, lowest free-energy state. Images 1.3a to i show an indication of such metastable states. Images a-b-c, d-e-f and g-h-i show three different samples with a Fe thickness of 2.3 atomic layers measured in zero field. The first images (a, d, g) have been measured right after the transfer to the microscope and show 3 very different domain patterns, although the samples were prepared under the same conditions. At constant temperature the patterns evolve with time until after typically one hour they become stationary. These stationary patterns are shown in imaged b, e and h. A common feature is that the domain size has increased (notice the scale bar), but the patterns are still very different. This is a clear sign that at least two of the three patterns are not at equilibrium. Upon heating the samples to $0.99\,T_C$ (c, f, i), the domain size decreases strongly and the patterns become totally equivalent in spite of their different history. This shows that at high enough temperatures the sample is indeed able to reach a true thermodynamic equilibrium.

Obviously, time plays a crucial role for the characteristics of the domain pattern. Figure j shows the magnetization $\overline{m}$ measured in response to an applied field. The blue curve shows $\overline{m}(T)$ for a sample that has been cooled in a constant applied magnetic field of 42 µT. The black curve shows the response if the field is increased from zero to 42 µT at fixed temperature. The red curve shows the amplitude of $\overline{m}$ in response to an AC-field that is switched between ±42 µT every 10 seconds. At high temperatures the 3 curves agree reasonably well but they diverge strongly as the temperature is lowered. This points to an increase of the relaxation time with decreasing T.

The relaxation time τ can be measured by following the evolution of $\overline{m}(t) \propto \exp\left(\frac{-t}{\tau}\right)$ after the applied field is switched off. Figure k shows the temperature dependence of τ fitted by a Vogel-Fulcher-law as indicated in the figure, with $T_0 \approx 0.85 T_C$, $b \approx 4$ and $\tau_\infty \approx 6 \cdot 10^{-11}$ s. A bare Arrhenius fit gives very unphysical parameters. This points to a non-Arrhenius temperature dependence of the relaxation times and is a further indication of glassy behaviour in this system.

2 Theoretical framework

2.1 Interactions in a ferromagnet

A ferromagnet is made up of localized or delocalized magnetic moments resulting in general from a combination of the orbital and spin moments of the electrons. These moments interact through different mechanisms whose interplay determines the global behaviour of the ferromagnet. In this section we give a very short, qualitative, introduction of the interactions relevant to our experimental system. For a more detailed account see e. g. the book by Aharoni [1].

2.1.1 Interaction with an external field

Classically, the potential energy of a magnetic dipole $\vec{m}$ in a magnetic field $\vec{B}_{\text{ext}}$ is given by [2]

$$E_{\text{Field}} = -\vec{m} \cdot \vec{B}_{\text{ext}} = -\vec{m} \cdot \mu_0 \vec{H}_{\text{ext}} \tag{2.1}$$

As is obvious, the external field tends to align the magnetic moment parallel to it. For a macroscopic sample the magnetization $\vec{M}$ is defined as the magnetic moment per unit volume. The energy of the magnetized sample can therefore be calculated simply by integration

$$E_{\text{Field}} = -\mu_0 \int_V \vec{M} \cdot \vec{H}_{\text{ext}} \, \mathrm{d}^3 V \,. \tag{2.2}$$

2.1.2 Exchange interaction

The Coulomb interaction and the Pauli exclusion principle lead to an effective spin-spin interaction between neighbouring atoms. This fact is reflected in the Heisenberg Hamiltonian [3]

$$\mathcal{H}_{\text{Heisenberg}} = -\sum_{i,j} J_{ij} \vec{S}_i \cdot \vec{S}_j \tag{2.3}$$

where $\vec{S}$ is the quantum mechanical spin operator. The exchange integral J_{ij} is finite only if the wave functions of the atoms at positions i and j overlap. In most cases the sum may therefore be restricted to the nearest neighbours. If $J > 0$ the exchange interaction favours parallel alignment of neighbouring spins and the system is ferro-magnetic whereas for $J < 0$ the exchange is anti-ferromagnetic, leading to preferred anti-parallel alignment of the spins $\vec{S}_i$ and $\vec{S}_j$. In the Heisenberg model the exchange is isotropic.

2.1.3 Anisotropy

Spin-orbit interaction between the atomic orbitals in the crystal lattice and the spins of the electrons breaks the rotational symmetry assumed in the Heisenberg model. However, the connection between the crystal structure of the magnetic material and its magneto-crystalline anisotropy is not straight forward. In the classical picture employed here we may introduce a phenomenological Hamiltonian [1, 4]

$$\mathcal{H}_{\mathrm{u}} = -K_u \left(\vec{M} \cdot \vec{e}_z \right)^2 = -K_u M^2 \cos^2 \vartheta \qquad (2.4)$$

to describe a uniaxial anisotropy preferring an alignment parallel to the z-axis. Similar expressions may be introduced for other symmetries of the anisotropy.

An additional source of anisotropy may come from surfaces and interfaces [5] that break the translational symmetry along their normal direction.

2.1.4 Dipolar interaction and shape anisotropy

When modelling a ferromagnet as localized magnetic moments arranged on a lattice, the magnetostatic interaction between the moments is in first order given by the dipole term. Classically, a microscopic magnetic dipole $\vec{m}$ at the origin produces a magnetic dipole field that is given by [2]

$$\vec{B}_{\mathrm{Dip}}(\vec{r}) = \frac{\mu_0}{4\pi} \frac{3\vec{n}(\vec{n} \cdot \vec{m}) - \vec{m}}{|\vec{r}|^3} \qquad (2.5)$$

except at the position of the dipole itself. This stray field originating from each magnetic moment in a macroscopic sample gives a contribution to the local magnetic field at the positions of all other moments, leading to an additional interaction between the moments of the sample. Since the dipole field falls off as $1/r^3$ this interaction is long-ranged. The angle dependent part $3\vec{n}(\vec{n} \cdot \vec{m})$ of the dipole interaction produces an anisotropy that depends on the macroscopic arrangement of the magnetic moments, i.e. the shape of the magnetic body. This shape anisotropy is responsible for the axial orientation of the magnetization in a rod or the planar orientation in a slab.

2.1.5 Domains and domain walls

Combining the above interactions qualitatively leads to the picture shown in Fig. 2.1. The ferromagnetic exchange interaction tends to align all spins parallel to each other (image a) along any direction. The anisotropies resulting from the crystal structure,

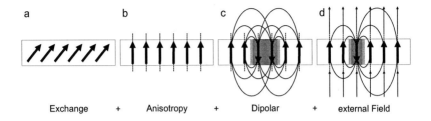

Exchange + Anisotropy + Dipolar + external Field

Figure 2.1: Effect of the different energy terms on a magnetized slab as discussed in the text.

interface and surface define an easy axis along which the spins preferentially align. Image b shows the case of a strong anisotropy perpendicular to the slab-like sample. A configuration with all spins — all local magnetic moments — parallel to each other and perpendicular to the sample plane leads however to a local magnetic field that is anti-parallel to the magnetic moments themselves. We are therefore in a situation of inherent frustration, where two interactions can not be minimized simultaneously: the exchange interaction together with the crystalline, surface and interface anisotropy prefers a state as in Fig. b, whereas the dipolar interaction favours a state with all spins lying in the plane. For weak perpendicular anisotropy the magnetostatic energy dominates and the sample will be magnetized homogeneously in the plane. In the case of strong perpendicular anisotropy drawn here and reflecting our experimental situation, the system can lower its total energy by breaking up into domains. Within each domain the constant magnetization minimizes the exchange and anisotropy parts of the total energy while the presence of oppositely magnetized domains lowers the dipolar energy. This situation is shown in image c. However, creating domains implies the creation of domain walls in which the magnetization has to rotate away from the easy axis (costing anisotropy energy) and neighbouring spins are no longer parallel (rising the exchange energy). The optimal magnetization distribution $\vec{M}(\vec{r})$ results form a delicate balance between the different energy terms. Adding an external field favours one magnetization direction with respect to the other, see image d.

2.2 Definition of the model

In order to approach our experimental situation we consider a slab of thickness d in the z-direction and lateral size 2Λ in the x-y-plane with $\Lambda \gg d$. We replace the discrete magnetic moments on a lattice by a continuous magnetization $\vec{M}(x, y, z)$. In the following we introduce a phenomenological model that fits our experimental situation of a perpendicularly magnetized thin film with mesoscopic domains.

2.2.1 Anisotropy

The slab is supposed to have a strong uniaxial anisotropy along the z-axis. Without worrying about its microscopic physical origin we consider the anisotropy by imposing that $\vec{M} \parallel \vec{e}_z$. This assumption is not fulfilled in the domain walls, where $\vec{M}$ rotates away from $\vec{e}_z$ but it is a good approximation as long as the domain walls are narrow compared to the domain size. Further we assume the magnetization $\vec{M}$ to be homogeneous along the z-direction and also within each domain. In this case we may write

$$\vec{M}(x,y,z) = \vec{e}_z M(x,y) = \vec{e}_z M_S m(\vec{\rho}) \tag{2.6}$$

where $\vec{\rho} = (x,y)$, $m(\vec{\rho}) = \pm 1$ and M_S is the saturation magnetization. Note that the field $m(\vec{\rho})$ describes only the geometry of the domain pattern. We can write

$$M_S = \frac{g\mu_B S}{a^3} \tag{2.7}$$

where $g\,S$ is the magnetic moment per atom in units of the Bohr-magneton μ_B and a^3 is the volume occupied by one atom.

2.2.2 Domain wall energy

If we assume the magnetization to be homogeneous inside the magnetic domains, then the exchange interaction is optimized except in the domain walls. The energy cost of a domain wall generally depends on its internal structure. Instead of writing a classical continuum equivalent of the exchange Hamiltonian (2.3) and calculating the associated energy we introduce the domain wall energy per unit wall area σ_w as a phenomenological parameter.

$$E_{wall} = l_{wall}\, d\, \sigma_w \tag{2.8}$$

where l_{wall} is the total length of all domain walls in the sample $(x$-$y)$-plane.

For Ising spins in a cubic lattice at zero temperature the only possible domain wall is a sharp step from $S_z = +1$ to -1 over one lattice constant a. In this case the cost of the domain wall per unit cell is $2J$ corresponding to one broken nearest neighbour bond, with J being the exchange constant. This value states an upper bound for the domain wall energy since such an atomically sharp domain wall is in principle always possible. More generally, the domain wall energy depends on the exchange constant J and the anisotropy K_u [1,4],

$$\sigma_w = \sqrt{2JK_u} \tag{2.9}$$

and the domain wall width w_{wall} is on the order of

$$w_{\text{wall}} = \sqrt{\frac{J}{2K_u}}. \tag{2.10}$$

2.2.3 External field

The interaction with an external magnetic field $\vec{H}_{ext} = H\vec{e}_z$ is given by (2.1)

$$E_{\text{Field}} = -\int_{-d/2}^{d/2} \int_{-\Lambda}^{\Lambda} \int_{-\Lambda}^{\Lambda} \mu_0 \vec{H}_{ext} \cdot \vec{M}(x,y,z)\, \mathrm{d}x\, \mathrm{d}y\, \mathrm{d}z \tag{2.11}$$

$$= -d\mu_0 H M_S \int_{-\Lambda}^{\Lambda} \int_{-\Lambda}^{\Lambda} m(x,y)\, \mathrm{d}x\, \mathrm{d}y \tag{2.12}$$

Where we have made use of the fact that the magnetization is homogeneous inside the domains and the domain wall width is small compared to the domain size as before. With this description we implicitly also assume that an external magnetic field may change the domain pattern but does not affect the magnetization inside the domains. This is justified as long as the Zeeman energy of the elementary magnetic moments is small compared to their exchange energy.

To make notation easier we redefine the field by

$$h = \mu_0 H M_S \tag{2.13}$$

and we introduce the geometrical magnetization

$$\overline{m} = \frac{1}{(2\Lambda)^2} \int_{-\Lambda}^{\Lambda} \int_{-\Lambda}^{\Lambda} m(x,y)\, \mathrm{d}x\, \mathrm{d}y\,. \tag{2.14}$$

The quantity $\overline{m}$ is the asymmetry of the area fractions $f_\uparrow$ and $f_\downarrow$ occupied by up ($\uparrow$) or down ($\downarrow$) magnetized domains respectively.

$$\overline{m} = \frac{f_\uparrow - f_\downarrow}{f_\uparrow + f_\downarrow}\,. \tag{2.15}$$

With these definitions the field-contribution to the energy becomes

$$E_{\text{Field}} = -d(2\Lambda)^2\, h\,\overline{m}\,. \tag{2.16}$$

2.2.4 Magnetostatic energy

The magnetostatic self-energy between magnetic (dipole) moments $\vec{m}_i$ on a lattice can be written as

$$E_{\text{MS}} = -\frac{\mu_0}{2} \sum_i \vec{m}_i \cdot \vec{H}_{\text{Demag}}(\vec{r}_i) \tag{2.17}$$

Where the demagnetizing field H_{Demag} is the magnetic field at the position of the moment $\vec{m}_i$, generated by the other magnetic moments of the sample. An obvious way to calculate it is as the superposition of dipole fields (2.5)

$$\mu_0 \vec{H}_{\text{Demag}}(\vec{r}_i) = \sum_{j,j\neq i} \frac{\mu_0}{4\pi} \frac{3\vec{n}(\vec{n}\cdot\vec{m}_j) - \vec{m}_j}{|\vec{r}_i - \vec{r}_j|^3} \tag{2.18}$$

with $\vec{n} = \frac{\vec{r}_j - \vec{r}_i}{|\vec{r}_j - \vec{r}_i|}$ [2]. For ultrathin (2-dimensional) magnetic films $\vec{n}$ lies always in the plane while $\vec{m}$ is perpendicular in the case of strong uniaxial anisotropy normal to the plane. In this case the anisotropic part of the dipolar energy involving $\vec{n} \cdot \vec{m}$ vanishes and we are left with

$$E_{Dip} = \frac{\mu_0}{8\pi} \sum_i \sum_{j, j \neq i} \frac{\vec{m}_i \cdot \vec{m}_j}{|\vec{r}_i - \vec{r}_j|^3} \tag{2.19}$$

where now the sample only extends in the x-y-plane.

When going to the continuum description, using this expression is dangerous because it becomes singular at $\vec{r}_i = \vec{r}_j$. One possible way to avoid the divergence is to introduce a cut off at some short length scale, e. g. the lattice constant or the domain wall width. An annoying side effect of this procedure is that the calculated quantities will then depend on the choice of this cut off. An alternative way is to compute the demagnetizing field via the magnetic charge $\rho_m = -\vec{\nabla} \cdot \vec{M}$ [6]. This is done in the appendix A.1, leading to the energy (A.14). By using (2.6), (2.7) and the definition

$$\lambda = \frac{\mu_0}{4\pi} M_S^2 \tag{2.20}$$

we obtain

$$E_{\mathrm{MS}} = \lambda \int_S m(\vec{\rho}) \int_S m(\vec{\rho}') \underbrace{\left(\frac{1}{|\vec{\rho} - \vec{\rho}'|} - \frac{1}{\sqrt{|\vec{\rho} - \vec{\rho}'|^2 + d^2}} \right)}_{V_d\left(|\vec{\rho} - \vec{\rho}'|\right)} \mathrm{d}^2 S' \, \mathrm{d}^2 S. \tag{2.21}$$

For large separations $r = |\vec{\rho}' - \vec{\rho}| \gg d$ the interaction potential $V_d(r)$ falls off like r^{-3}, see eq. (A.21), whereas for short distances $r \ll d$ it diverges as r^{-1} and is therefore integrable in 2D. Since $V_d > 0$, the magnetostatic interaction favours an anti-parallel arrangement of the magnetization on a larger scale.

2.2.5 Total energy

By collecting the energy terms we arrive at the total energy of a magnetic slab with strong perpendicular anisotropy:

$$E = l_{wall} \, d \, \sigma_w - d(2\Lambda)^2 \, h \, \overline{m} + \lambda \iint m(\vec{\rho}) \, m(\vec{\rho}') \, V_d\left(|\vec{\rho} - \vec{\rho}'|\right) \, \mathrm{d}^2 \rho' \, \mathrm{d}^2 \rho \tag{2.22}$$

In zero external field the first energy term favours a state without domain walls, i.e. the sample being in a homogeneously magnetized state with all magnetic moments parallel. However, in such a state the magnetostatic energy as given by the last term is maximal. The competition between these two energies leads to the formation of a domain pattern that is characterized by the ratio of the coupling constants

$$\frac{C_{\text{Wall}}}{C_{\text{MS}}} = \frac{\sigma_w}{\lambda} \tag{2.23}$$

and the thickness of the film d. The model (2.22) is exact for Ising spins (infinite anisotropy) at zero temperature, where the domain walls are atomically thin, and the magnetization inside the domains equals $\pm \frac{g\mu_B S}{a^3}$. We expect it to be approximately valid also for weaker perpendicular anisotropy and at finite temperature as long as the underlying assumptions are fulfilled. We repeat them explicitly:

- The domain walls are narrow compared to the domain size.

- The local magnetization in each domain is homogeneous and pointing along $\pm \vec{e}_z$.

- The absolute value of the local magnetization is the same for up- and down magnetized domains.

These conditions are justified by mean-field calculations [7] and are confirmed experimentally up to rather high temperatures, see section 4.5.

2.3 Regular arrays of domains

2.3.1 Stripe lattice in zero field

The ground state of a magnetic sample as described in the previous section in absence of a magnetic field is expected [8–11] and observed [12–15] to consist of magnetic stripe domains of alternating magnetization for both, ultra-thin and thick magnetic films. To get a handle on what 'ultra-thin' and 'thick' mean, we investigate a perfect stripe pattern of period $2L$, i. e. stripe width L. The energy density of such a system is given by equation (A.75) derived in appendix A.2.

$$e = \frac{\sigma_w}{L} + \lambda \frac{16}{\pi^2} \frac{L}{d} \left[\frac{7}{8} \zeta(3) - \frac{1}{2} Li_3 \left(e^{-\frac{\pi d}{L}} \right) + \frac{1}{2} Li_3 \left(-e^{-\frac{\pi d}{L}} \right) \right] , \tag{2.24}$$

where $Li_3(x) = \sum_{n=1}^{\infty} n^{-3} x^n$ is the trilogarithm. For small $\frac{d}{L}$ we can use the expansion (A.76) to write

$$-\frac{1}{2} Li_3 \left(e^{-\frac{\pi d}{L}} \right) + \frac{1}{2} Li_3 \left(-e^{-\frac{\pi d}{L}} \right)$$
$$= -\frac{7}{8} \zeta(3) + \frac{\pi^3}{8} \frac{d}{L} - \frac{\pi^2}{4} \left(\ln \left(\frac{2L}{\pi d} \right) + \frac{3}{2} \right) \left(\frac{d}{L} \right)^2 + \mathcal{O} \left(\frac{d^4}{L^4} \right) \tag{2.25}$$

11

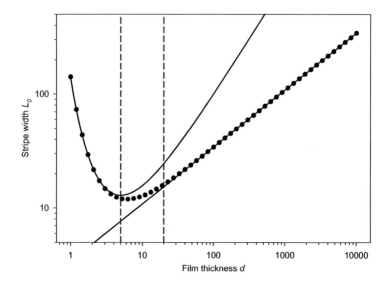

Figure 2.2: Ground state stripe width L_0 vs. film thickness d for $\lambda = 0.1$ and $\sigma_w = 2$ presented in a log-log-plot. The dots correspond to the numerical results obtained from minimizing the energy (2.24) and the black lines show the analytical expressions (2.27) and (2.29) with their respective ranges of validity delimited by the dashed lines.

and obtain for the energy density

$$e = \frac{\sigma_w}{L} + 2\pi\lambda - \frac{4\lambda d}{L}\left(\ln\left(\frac{2L}{\pi d}\right) + \frac{3}{2}\right). \tag{2.26}$$

Minimizing gives the equilibrium stripe width in the ground state

$$L_0 = d\frac{\pi}{2}e^{\frac{\sigma_w}{4\lambda d} - \frac{1}{2}} \qquad \left(d < \frac{\sigma_w}{4\lambda}\right). \tag{2.27}$$

For large $\frac{d}{L}$, the Li_3-terms in (2.24) can be neglected. The total energy density becomes

$$e = \frac{\sigma_w}{L} + \frac{14\lambda\zeta(3)}{\pi^2}\frac{L}{d} \tag{2.28}$$

and the corresponding L_0 reads

$$L_0 = \sqrt{\frac{\sigma_w d}{\lambda}\frac{\pi^2}{14\zeta(3)}} = d\frac{\pi}{2}\sqrt{\frac{\sigma_w}{4d\lambda}}\sqrt{\frac{16}{14\zeta(3)}} \qquad \left(d > \frac{\sigma_w}{\lambda}\right). \tag{2.29}$$

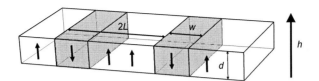

Figure 2.3: Definition of the stripe lattice parameters: w - width of the minority stripes, $2L$ - periodicity of the lattice, d - thickness of the film.

From Fig. 2.2 we can distinguish two regimes for the behaviour of L_0 vs. d: In (ultra-) thin films with $d < \frac{\sigma_w}{4\lambda}$ the stripe width L_0 increases exponentially with $\frac{\sigma_w}{4\lambda d}$ according to eq. (2.27). As a consequence even relatively small changes in σ_w or λ have a drastic effect on the domain size. Thick films are more robust, as for $d > \frac{\sigma_w}{\lambda}$ (2.29), the stripe width increases only as $\sqrt{\frac{\sigma_w d}{\lambda}}$. Notice that also the systems here referred to as 'thick' have $d \ll \Lambda$. Since our experimental situation deals with ultra-thin films and the case of thick films has been studied extensively in the literature [8,12,16,17] we concentrate in the following on the ultra-thin case.

2.3.2 Stripes in an applied field

In the presence of a magnetic field equation (2.24) is no longer valid as the stripes with the magnetization parallel to the external field will be favoured. We define w as the width of the smaller stripes and $2L$ as the period of the stripe lattice, see Fig. 2.3. For the ultra-thin case an analytical form of the dipolar energy can be derived, see appendix A.2, equation (A.123):

$$e_{d/l} = \frac{\sigma_w}{L} + \frac{w}{L}h - \frac{4\lambda d}{L}\left[\ln\left(\frac{2L}{\pi d}\sin\left(\frac{\pi w}{2L}\right)\right) + \frac{3}{2}\right] \qquad (2.30)$$

An almost equivalent expression has already been found by Kashuba and Pokrovsky [18] with a slightly different treatment of the dipolar interaction, hence their expression has no term $+\frac{3}{2}$ after the logarithm. By minimizing this energy we obtain (see appendix A.2)

$$L = L_0\frac{1}{\sqrt{1 - \frac{h^2}{h_{C,s}^2}}} \qquad (2.31)$$

with L_0 as given by (2.27) and

$$h_{C,s} = 2\pi\lambda d\frac{1}{L_0} = 4\lambda e^{\frac{1}{2}}e^{-\frac{\sigma_w}{4\lambda d}} . \qquad (2.32)$$

The minority stripe width is given by

$$w = \frac{2}{\pi} L \arcsin\left(\frac{L_0}{L}\right) \tag{2.33}$$

which is again in agreement with the results of Kashuba and Pokrovsky except for a factor 2 as discussed in the context of (A.123). We can check (2.33) and (2.31) for the limiting cases:

i) $h = 0$ gives trivially $L = L_0$ and $w = L_0$ as expected.

ii) For $h \to h_{C,S}$ we have $L \to \infty$. Taking the limit in (2.33) gives

$$w_C = \lim_{L \to \infty} \frac{2L}{\pi} \arcsin\left(\frac{L_0}{L}\right) = \lim_{L \to \infty} \frac{2L}{\pi} \frac{L_0}{L} = \frac{2}{\pi} L_0 \tag{2.34}$$

From (2.33) and (2.31) we can calculate the magnetization $\overline{m}(h)$:

$$\overline{m}(h) = 1 - \frac{w}{L} = 1 - \frac{2}{\pi} \arcsin\left(\frac{L_0}{L}\right) \tag{2.35}$$

The susceptibility can then be calculated as

$$\chi(h) = \frac{\mathrm{d}\overline{m}}{\mathrm{d}h} = \left(\frac{\partial \overline{m}}{\partial \frac{L_0}{L}}\right) \frac{-L_0}{L^2} \left(\frac{\partial L}{\partial h}\right) \tag{2.36}$$

$$= \frac{2}{\pi} \frac{1}{h_{C,S}} \frac{1}{\sqrt{1 - \frac{h^2}{h_{C,S}^2}}} = \frac{1}{\pi^2 \lambda d} L \tag{2.37}$$

and the initial susceptibility is

$$\chi_0 = \chi(0) = \frac{1}{\pi^2 \lambda d} L_0 = \frac{2}{\pi} \frac{1}{h_{C,S}}. \tag{2.38}$$

2.3.3 Hexagonal bubble lattice

As has been shown earlier for both the thick [19, 20] and thin [21] film case, with increasing magnetization $\overline{m}$ at some point a hexagonal lattice of inverted circular domains (bubbles) has lower energy than the stripe pattern. We consider circular bubbles of radius R aligned on a hexagonal lattice of spacing L. The local contributions to the energy density for this case read

$$e_{\text{Field}} = 2\frac{\pi R^2}{L^2 \frac{\sqrt{3}}{2}} h = \frac{4\pi R^2}{\sqrt{3} L^2} h \tag{2.39}$$

$$e_{\text{Wall}} = \frac{2\pi R \sigma_w}{L^2 \frac{\sqrt{3}}{2}} = \frac{4\pi R \sigma_w}{\sqrt{3} L^2}. \tag{2.40}$$

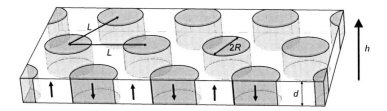

Figure 2.4: Definition of the bubble lattice parameters: L - lattice spacing, R - bubble radius, d - thickness of the film. In our experiments $R \gg d$, R/d is larger than 1000 and the bubble domains are rather disks than cylinders.

The magnetostatic energy can not be expressed analytically. An expansion for small $\frac{d}{R}$, equation (A.188) is calculated in appendix A.3 and the total energy density reads:

$$
\begin{aligned}
e_{Tot} &= \frac{4\pi R^2}{\sqrt{3}L^2} h + \frac{4\pi R\sigma_w}{\sqrt{3}L^2} + \frac{4\pi\lambda dR}{L^2\sqrt{3}} \left[-4\ln\left(\frac{8R}{d\sqrt{e}}\right) + 6\pi \sum_{k=0}^{\infty} a_{2k} S_{2k} \left(\frac{R}{L}\right)^{3+2k} \right] \\
&= \frac{4\pi R}{\sqrt{3}L^2} \left[\sigma_w + Rh - 4\lambda d\ln\left(\frac{8R}{d\sqrt{e}}\right) + 6\pi\lambda d \sum_{k=0}^{\infty} a_{2k} S_{2k} \left(\frac{R}{L}\right)^{3+2k} \right] \quad (2.41)
\end{aligned}
$$

Since the coefficients a_{2k} and S_{2k} are independent of both R and L, their values need to be calculated only once and (2.41) can efficiently be evaluated numerically.

As becomes evident in figure 2.7, the transition to the saturated state occurs by a divergence of L while the bubble radius R stays finite. We can therefore calculate analytically the critical field h_C at which the uniform state becomes lowest in energy from (2.41) by dropping the sum-term and minimizing the remaining energy of a single bubble with respect to R. We find, see section 2.4.2,

$$
h_C = 32\lambda \exp\left(-\frac{\sigma_w}{4\lambda d} - \frac{3}{2}\right) = \frac{16\pi\lambda d}{e^2} \frac{1}{L_0} = \frac{8}{e^2} h_{C,S} \quad (2.42)
$$

2.3.4 Stripes or bubbles - a simple estimate

Based on the total energy (2.22) we can state a simplified energy density per unit volume

$$
e = e_{\text{Exchange}} + e_{\text{Field}} + e_{\text{Dipolar}} = l_{\text{wall}}(L,\overline{m})\sigma_w - h\overline{m} + e_\lambda(L,\overline{m}) \quad (2.43)
$$

We have seen that the competition between dipolar and exchange energy leads to modulation of the system on a mesoscopic scale L. The applied field h favours one magnetization direction, leading to $\overline{m} \neq 0$. In this case, the competition between the applied field and the dipolar term determines $\overline{m}$ or equivalently $f_\downarrow = \frac{1}{2}(1 - \overline{m})$, the

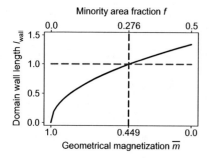

Figure 2.5: Estimate of the cross-over from the stripe to the bubble lattice as a function of the geometrical magnetization $\overline{m}$. The domain wall lengths are plotted for the stripes (red dashed) and bubbles (blue continuous). The black dashed line marks the cross-over from stripes to bubbles.

area fraction occupied by the minority domains. Because the dipolar interaction is weak and long-ranged, we can think that it is insensitive to the exact geometry of the domain pattern. Assuming that we have a certain f and L established, the field- and dipolar terms of the energy are approximately optimized and the only term left is the exchange energy. The exchange energy depends only on the domain wall length per unit area l_{wall} and minimizing the energy at a given L and f corresponds therefore to minimizing l_{wall}. We can compare the energies of a stripe lattice with period L_S and a bubble lattice with period L_B by noticing that

$$\text{Stripes} \quad l_{\text{wall,S}} = \frac{2}{L_S} \qquad\qquad f_S = \frac{w}{L_S} \qquad (2.44)$$

$$\text{Bubbles} \quad l_{\text{wall,B}} = \frac{4\pi R}{\sqrt{3}L_B{}^2} \qquad\qquad f_B = \frac{2\pi R^2}{\sqrt{3}L_B{}^2}. \qquad (2.45)$$

While for the stripes the domain wall length is independent of f, for the bubbles we have

$$l_{\text{wall,B}} = \frac{2}{L_B}\sqrt{\frac{2\pi}{\sqrt{3}}}\sqrt{f_B} \qquad (2.46)$$

If we set now $L_S = L_B = L$ we can plot $l_{\text{wall}}(f)\frac{L}{2}$ for the two lattices, see Fig. 2.5. From this simple argument we expect the cross-over from stripes to bubbles for $f \approx \frac{\sqrt{3}}{2\pi} = 0.276$ or equivalently $\overline{m} \approx \pm 0.45$.

2.3.5 Comparing the stripe and the bubble lattice

By combining the results from the previous sections we can now investigate the ground-state phase diagram of a magnetic film in a magnetic field more closely. The respective energies need to be evaluated carefully since they differ only by a few percent over the entire range from $h = 0$ to h_C.

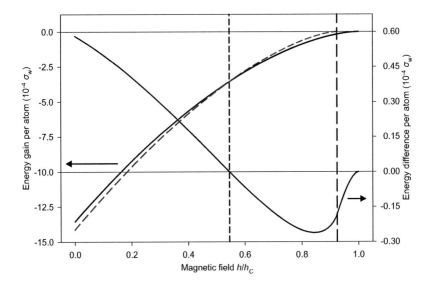

Figure 2.6: Total energy gain per atom for the optimized pattern of stripes (red dashed, left scale) and bubbles (blue continuous, left scale) with respect to the uniform state in units of $\sigma_w \cdot 10^{-4}$ for λ=0.1, σ_w=2 and d=1. The black curve is the difference of the two energies (right scale) in the same units. The vertical black dashed lines correspond to the crossover field h_{SB} and the critical field for the stripes $h_{C,S}$ respectively.

The energies of the stripe lattice (2.30) and the bubble lattice (2.41) are compared in Fig. 2.6 for the parameter values $d = 1, \sigma_w = 2, \lambda = 0.1$. We find numerically that the stripe array is the lowest energy state up to a cross-over field h_{SB} above which the hexagonal bubble lattice is favourable until at the critical field $h_C = h_{C,B}$ the uniform state has the lowest energy. The numerical value obtained for $d = 1$ is

$$h_{\mathrm{SB}} = 0.545(3) \cdot h_C \qquad (2.47)$$

and is independent of the ratio $\frac{\sigma_w}{\lambda}$ as long as the thin-film approximation $\left(d < \frac{\sigma_w}{4\lambda}\right)$ holds. Figure 2.7 shows the variation of L_S, L_B, R and w over the full magnetic field range from $h = 0$ to h_C. We observe that at low fields the stripe period L_S is constant while the minority stripe width w decreases linearly with the magnetic field. This gives rise to an initially constant susceptibility, $\chi = \frac{d\overline{m}}{dh} = $ const. as calculated in eq. (2.38). The complete magnetization vs. field curves of the bubble and the stripe lattice are plotted in figure 2.8. At zero field, the magnetization of the stripe lattice is zero as expected. In contrast, the magnetization $\overline{m}$ of the lowest energy bubble lattice

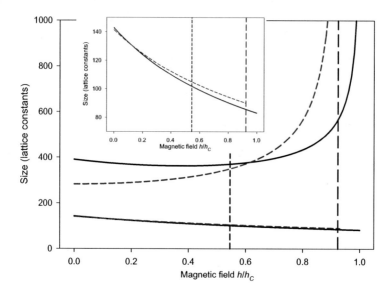

Figure 2.7: Domain sizes for optimized stripe (red dashed) and bubble (blue continuous) arrays for the parameters (d=1, σ_w=2, λ=0.1). The upper curves are the lattice periods $L_S = 2L$ and L_B, the lower curves the characteristic domain sizes w and R. Since the latter are very close over the entire magnetic field range, they are displayed in the inset for better comparison. The vertical black dashed lines correspond to the crossover field h_{SB} and the critical field for the stripes $h_{C,S}$ respectively.

at $h = 0$ is finite. Also for the bubble lattice the initial susceptibility is linear. For $d = 1, \sigma_w = 2, \lambda = 0.1$ we find numerically

$$\overline{m}(h) = \overline{m}_0 + \chi_B(0)\, h \qquad \text{with} \qquad \overline{m}_0 = 0.0306 \quad \text{and} \quad \chi_B(0) = 0.757\frac{1}{h_C}. \quad (2.48)$$

At the cross-over field h_{SB} we find the magnetization value for the bubble lattice $\overline{m} = 0.454$ being higher than for the stripe lattice ($\overline{m} = 0.402$) in exact agreement with the values calculated by Ng and Vanderbilt [21][1] With increasing field, a jump in the magnetization should therefore occur as the pattern transforms from a stripe lattice to a bubble lattice.

Figure 2.9a and b show the optimized bubble- and stripe lattices superimposed at $h = h_{SB}$ for two orientations of the stripe lattice. There is a small lattice mismatch

---

[1]Note that the area fraction f occupied by the up-domains used by Ng and Vanderbilt is equivalent to the $\overline{m}$ used here: $\overline{m} = 2f - 1$.

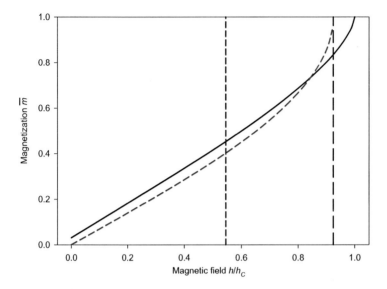

Figure 2.8: Magnetization $\overline{m}$ for optimized stripe (red dashed) and bubble (blue continuous) arrays as a function of magnetic field. The vertical black dashed lines correspond to the crossover field h_{SB} and the critical field for the stripes $h_{C,S}$ respectively. Note that at the cross-over field the magnetization of the stripes is lower than the one of the bubbles.

between the two. Figure c shows a close-up of the lattice constants in the cross over region. One identifies that the bubble row separation (blue dotted curve) matches the stripe spacing (red dashed curve) at $h \approx 0.4\, h_C$ while the bubble lattice spacing (blue continuous curve) matches the stripe lattice at $h \approx 0.62\, h_C$. In an experiment one may therefore expect the cross-over to occur via some intermediate state somewhere between 0.4 and $0.6\, h_C$. The transformation of the stripe lattice to the bubble lattice requires the stripes to break apart and the reverse process requires bubbles to merge into stripes. Such fission or fusion implies the crossing of an energy barrier as we will discuss in more detail in section 2.6. If the fluctuations in the system are too weak (i. e. the temperature is too low) stripe fission and fusion are prohibited and it is impossible for the system to reach the equilibrium state. As a consequence, in such a case the bubble state will most likely not be observed and the stripes will persist up to h_C and beyond.

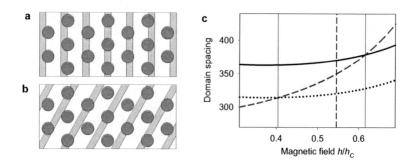

Figure 2.9: Optimal stripe and bubble lattices at the cross over field. **a** Bubble rows and stripes aligned – the bubble row spacing is slightly smaller than the period stripe lattice. **b** Stripe lattice rotated by $30°$ – the bubble lattice constant is slightly larger than the stripe lattice. **c** Lattice spacing for the bubble lattice L_B (blue continuous), stripe period L_S (red dashed) and bubble row spacing $\frac{\sqrt{3}}{2}L_B$ (blue dotted) vs. applied field. The solid vertical lines indicate where the lattice constants match, the dashed vertical line indicates the equilibrium cross-over field H_{SB}.

2.4 Isolated domains

2.4.1 Single stripe domain

We have seen that the stripe lattice becomes more and more diluted as h approaches $h_{C,S}$ and the transition to the bubble lattice will be suppressed over most of the temperature range. For this reason, we will investigate the metastability of the stripe domains close to and beyond $h_{C,S}$ in more detail. With this purpose we consider the energy of one reversed stripe domain in an otherwise homogeneous film. The corresponding energy per unit stripe length is given in the appendix, eq. (A.137):

$$E = 2\sigma_w + 2wh - 8\lambda d \left[\ln\left(\frac{w}{d}\right) + \frac{3}{2} \right] \tag{2.49}$$

This energy is plotted in figure 2.10a for d=1, σ_w=2, λ=0.1.
The minimum of the energy is found for

$$\frac{\partial E}{\partial w} = 2h - \frac{8\lambda d}{w} = 0 \quad \Longrightarrow \quad w^* = \frac{4\lambda d}{h}. \tag{2.50}$$

For $h > h_{C,S}$ as given by (2.32) the stripe domain is metastable but the local energy minimum defined by the above equation is always at $w^* > 0$ and the energy barrier to $w = 0$ never collapses. Of course for $w \approx a = d$ the continuum approximation breaks down, the stripe can certainly not become smaller than the lattice constant. Using

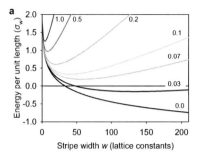
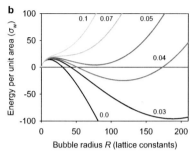

Figure 2.10: a Energy per unit length of a single reversed stripe in a homogeneous background as a function of its width w for different applied magnetic fields h. The magnetic field values are indicated in units of λ next to every curve (σ_w=2, λ=0.1, d=1). The critical field at which the stripe energy is equal to zero is $h_{C,S} = 0.0443 \cdot \lambda$ according to (A.133). **b** Energy per unit area of a single reversed bubble as a function of its radius R.

this condition we can state an upper bound for the collapse field by setting $w^* = a = d$ in (2.50) and solving for h. We obtain

$$h_{\text{collapse,S}} \leq 4\lambda. \tag{2.51}$$

For the parameters σ_w=2, λ=0.1 and d=1 we have that the bound for the collapse field found in this way exceeds the equilibrium saturation field h_C (2.42) by two orders of magnitude. A more refined estimate is discussed in section 2.6.1.

2.4.2 Single bubble domain

We consider the gain in energy by inserting one single reversed bubble of radius R in an otherwise homogeneous thin ($d \ll R$) film. An expression for this energy has been given by Thiele [22], we adapt it to our nomenclature:

$$E = 2\pi R d \left(\sigma_w - 4\lambda d \ln\left(\frac{8R}{d\sqrt{e}}\right) + Rh \right) \tag{2.52}$$

This energy is plotted in Fig. 2.10b. We can distinguish the following cases:

i) $h \leq 0$: The state with $R = 0$ is metastable, the formation of a bubble is hindered by an energy barrier. Once the bubble radius is large enough the energy is monotonously decreasing with growing R. Of course, for $h < 0$ the favoured state is the completely reversed state, i. e. the bubble filling the entire sample.

ii) $0 < h < h_{C,B}$: The energy has a global minimum at a finite radius R^* that depends on the applied field. The state at $R = 0$ is still metastable

iii) $h_{C,B} < h < h_{collapse,B}$: there are two stable configurations: one with $R^* > 0$ that is meta stable and the one with $R = 0$ that is now the state of lowest energy.

iv) $h > h_{collapse,B}$: the energy is monotonously increasing with R and the bubble collapses inevitably to the only stable state $R = 0$, corresponding to the sample being magnetized completely parallel to the applied field.

The following conditions lead to the critical fields:

$$h = h_{C,B} \quad \Leftrightarrow \quad \frac{\partial E(R,h)}{\partial R} = E(R,h) = 0 \tag{2.53}$$

$$h = h_{collapse,B} \quad \Leftrightarrow \quad \frac{\partial^2 E(R,h)}{\partial R^2} = \frac{\partial E(R,h)}{\partial R} = 0 \tag{2.54}$$

Giving their values:

$$h_{C,B} = 32\lambda \exp\left(-\frac{\sigma_w}{4\lambda d} - \frac{3}{2}\right) = \frac{16\pi\lambda d}{e^2}\frac{1}{L_0} =: h_C \tag{2.55}$$

$$h_{collapse,B} = 16\lambda \exp\left(-\frac{\sigma_w}{4\lambda d} - \frac{1}{2}\right) = \frac{8\pi\lambda d}{e}\frac{1}{L_0} = \frac{e}{2}h_C \tag{2.56}$$

where L_0 is the stripe width in zero field as given by (2.27). The field $h_{C,B}$ as stated above is the (equilibrium) saturation field of the sample: for fields higher than $h_{C,B}$ the uniform, saturated, state is lowest in energy as anticipated in section 2.3.3. We will refer to it simply as h_C. The calculations of the field values are given in appendix A.3.3.

The radius of the isolated bubble for these two field values is:

$$R_C = \frac{d}{8}e^{\frac{\sigma_w}{4\lambda d} + \frac{3}{2}} = \frac{e^2}{8}w_c = \frac{e^2}{4\pi}L_0 \tag{2.57}$$

$$R_{collapse} = \frac{d}{8}e^{\frac{\sigma_w}{4\lambda d} + \frac{1}{2}} = \frac{e}{8}w_c = \frac{e}{4\pi}L_0 \tag{2.58}$$

2.5 Scaling properties

In this section we identify two scaling laws present in the zero-temperature results obtained in the previous sections and tentatively extend them to finite temperatures. The ground state properties of the sample are governed by two parameters: The film thickness d and the ratio of wall energy to the dipolar coupling constant, σ_w/λ. Although we have not considered temperature in any of the preceding derivations, we may still expect the results for the energy to be valid also at finite temperature as long as the underlying assumptions stated in section 2.2.5 are fulfilled and the temperature dependence of the free energy may be taken into account by introducing effective temperature dependent parameters $\sigma_w(T)$ and $\lambda(T)$.

2.5.1 Scaling properties at zero temperature

In the previous sections we have found that L_0, the stripe width in zero magnetic field, constitutes a fundamental length in the system. As has been shown in (2.27) for the ultrathin film case relevant here, this length relates to the model parameters via

$$L_0 = d \, \frac{\pi}{2\sqrt{e}} \, \exp\left(\frac{\sigma_w}{4\lambda d}\right) . \tag{2.59}$$

We have seen that all length scales, L_S, L_B, R, w scale with L_0. On the other hand, we have also shown that the magnetic fields $h_{SB}, h_{C,S}, h_{\text{collapse,B}}$ are proportional to h_C. From (2.55) we see that the magnetic fields scale with $\frac{\lambda}{L_0}$ since for a given sample its thickness d is a constant. Using the definition of λ we can relate the dipolar coupling to

$$\lambda = \frac{\mu_0}{4\pi} \left(\frac{g\mu_B S}{a^3}\right)^2 = \frac{\mu_0}{4\pi} (M_S)^2 \tag{2.60}$$

If we remember that

$$h = \mu_0 \frac{g\mu_B S}{a^3} H = \mu_0 M_S H = \frac{16\pi d}{e^2} \frac{\lambda}{L_0} = \frac{4\mu_0 d}{e^2} \frac{(M_S)^2}{L_0} \tag{2.61}$$

We see that the saturation field — and with it all critical fields — scales as [18,23]

$$\mu_0 H_C = \frac{4\mu_0 d}{e^2} \frac{M_S}{L_0} \qquad \text{and} \qquad H_C \propto \frac{M_S}{L_0} \tag{2.62}$$

In section 2.3.2 we have seen that the susceptibility $\chi = \frac{\partial \overline{m}}{\partial h}$, eq. (2.38), is constant over a wide range of applied magnetic fields, see figure 2.8. We can therefore also identify an approximate scaling law for $\overline{m}$

$$\overline{m} = \chi_0 h = \frac{1}{\pi^2 \lambda d} L_0 \, h \propto \frac{L_0}{M_S^2} M_S H = \frac{L_0 H}{M_S} . \tag{2.63}$$

2.5.2 Finite temperature

The behaviour of the striped system in zero field at finite temperature has been studied previously in great detail [6]. As has been shown recently also in calculations [7], the influence of temperature can be summarized in three effects:

i) A broadening of the domain wall with increasing temperature.

ii) A reduction of the saturation magnetization or the magnetization inside a domain with increasing temperature [4].

iii) A drastic decrease of the the zero field stripe width $L_0(T)$ at intermediate temperatures and a less pronounced decrease at high temperatures.

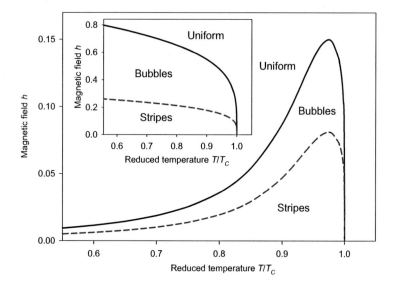

Figure 2.11: Phase diagram in H-T-space predicted using the scaling property (2.62) for $\lambda_0 = 0.1$, $\sigma_{w0} = 2$, $d = 1$ in combination with the expressions (2.64) and (2.65). The inset shows the phase diagram obtained from the scaling $H_c \propto M_S(T)$ as it is expected for thick films.

Experimentally and also theoretically, the following expressions were found for the temperature dependence of the domain width and the magnetization close to T_C [18, 24]:

$$L_0(T) = L_C + L_1(1 - T/T_C)^2 \quad \text{with} \quad L_C = \frac{J}{\lambda d} \tag{2.64}$$

and

$$M_S(T) = M_0(1 - T/T_C)^{\beta} \tag{2.65}$$

where we have introduced the reduced temperature $\frac{T}{T_C}$. Using these expressions together with the relation (2.62) we can predict the phase diagram in the T-H-plane, see Fig. 2.11 by plotting the transition line $H_C(T)$ and $H_{SB}(T) = 0.55H_C(T)$, see equation (2.47). From the graph it is evident that – except for the highest temperatures – the transition lines curve upwards, i.e. the uniform saturated state occurs at lower temperature than the modulated phase. This behaviour has been predicted by Abanov *et al.* [23] for ultrathin films in contrast to the phase diagram calculated by

a

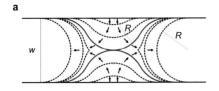

b

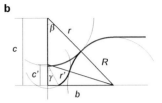

Figure 2.12: a Illustration of the fission of a long stripe as discussed in the text. The solid red line marks the critical state at the maximum barrier height and the dashed lines mark intermediate states. **b** Geometry of the domain boundary for the fission as modeled here, see text.

Garel and Doniach for thick films [8] (inset). Notice however, that close to T_C our model is expected to break down because the domain size and the domain wall width become comparable, see section 4.5 and [7].

The upward-bending of the transition lines predicts inverse symmetry breaking. At low temperature, in a constant magnetic field of appropriate magnitude, a uniform, saturated state is expected. This state is completely isotropic in the x-y-plane. As the temperature is increased, a transition to a bubble lattice occurs. This state has only discrete translational and rotational symmetries, i. e. the system's symmetry is *lowered* by *heating*. A further reduction in symmetry occurs at the transition from the bubble to stripe lattice, where the rotational symmetry is only 2-fold.

2.6 Domain transformations

2.6.1 Splitting a single stripe

In section 2.4 we have seen that for $h > h_{C,S}$ an isolated stripe becomes metastable while an isolated bubble is stable up to $h = h_C$. Above $h_{C,S}$ the stripe will therefore contract along its length to form a bubble. If we assume that the stripe ends are fixed by some pinning center, the stripe can not collapse by contraction. As we have seen, the field needed to compress the stripe laterally over its entire length is very high. Let us consider now a different process: we want to estimate the energy barrier for compressing the stripe punctually, i. e. for breaking the stripe apart at some arbitrary point along its length. To estimate the barrier height, we model the breaking as shown in Fig. 2.12a: A symmetric constriction of radius $r = w/2$ is inserted in a stripe of width w. The energy is given by

$$E(b,c) = l(b,c)\sigma_w + A(b,c)h_{eff} \qquad (2.66)$$

Where b and c parametrize the path from the initial state to the final state, l is the change in domain wall length and A the change in domain area. The demagnetizing

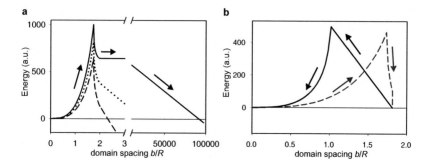

Figure 2.13: Energy barrier for **(a)** the fission of an infinite stripe and **(b)** fusion of two bubble domains (blue continuous curve) or fission of one elongated domain to two bubbles (red dashed curve) in a lattice. The path taken in a process is indicated by the arrows and the b-parameter is as defined in the text. Notice the break in the b-scale of fig. a.

field resulting from the dipolar interaction with the surrounding sample and the external field are combined into an effective field h_{eff} which is assumed to be homogeneous in the small sample volume considered for the process. The geometry for one quadrant of the constriction is sketched in Fig. 2.12b.

Geometrically, the fission is described by a circle of radius r moving across the stripe from both sides. The distance of this circle centre from the centreline of the domain we refer to as c. In the final state, both domains will be terminated by a semicircle of radius $R = w/2$. We use the distance from the center of this circle to the center of the constriction as the second parameter b. With these definitions the change in boundary length reads

$$l(b,c) = 4\left(-b + R\beta + r\beta - r\gamma\right) \tag{2.67}$$

with

$$\beta = \arctan\left(\frac{b}{c}\right) \quad , \quad \gamma = \begin{cases} \arccos\left(\frac{c}{r}\right) & c \leq r \\ 0 & c > r \end{cases} \quad \text{and} \quad r = \sqrt{b^2 + c^2} - R \tag{2.68}$$

as defined in Fig. 2.12b. The area difference is

$$A(b,c) = 4\left(-bR + \frac{bc}{2} + \frac{R^2\beta}{2} - \frac{(\beta-\gamma)r^2}{2} - \frac{cr\sin(\gamma)}{2}\right). \tag{2.69}$$

For an infinite stripe we choose $r = R$ and thus $c = \sqrt{4R^2 - b^2}$. The energy barrier for this case is plotted in Fig. 2.13a for three different values of the external field h_{ext}. The dipolar field contribution to h_{eff} is given by the condition that at the critical field $h_{C,S}$ the state with the stripe and the uniform state (corresponding to the completely

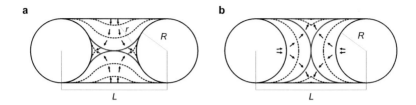

Figure 2.14: Illustrations of **(a)** domain fission and **(b)** domain fusion. The solid black lines mark the initial and the final states, the solid red line marks the critical state at the maximum barrier height and the dashed lines mark intermediate states.

separated state, $b = \Lambda$) have the same energy (solid curve). From this condition we obtain $h_{dip} \approx -5h_C$. The dotted curve in Fig. 2.13a is for $h_{ext} = 2h_C$ and the dashed curve for $h_{ext} = 3h_C$. As expected, the barrier height decreases with increasing h_{ext} and for an external field h_{ext} larger than about $7h_C$ the energy barrier vanishes and the stripe will split spontaneously. Note that this value is less than $1/10$ of the collapse field calculated in (2.51) but is still considerably larger than the (equilibrium-) saturation field.

2.6.2 Domain fusion and fission in a lattice

As we have seen in section 2.3.5, the bubble lattice has a lower energy at high fields whereas the stripes are favourable at low fields. Therefore, at some point a cross-over from the stripes to the bubble lattice has to take place. However, if the lattice constant remains the same, this involves a large number of domain splitting / domain merging events. The situation is somewhat different from the isolated stripe, see Fig. 2.14a. The separated domains can not separate infinitely but only up to $b = L$ where L is the domain center spacing of the pattern. The fission is assumed to occur in the same way as before except that we set $r = min(\sqrt{b^2 + c^2} - R, R)$ and $b \in [0, L]$ as boundary conditions. For the fusion we assume that the domains grow along their connecting line until the domain walls touch. Then the 360° wall between the domains collapses and one elongated domain results as the final state. The dipolar field is chosen such that the elongated domain and the two isolated bubbles have the same energy at the cross-over field $h_{ext} = h_{SB}$. We obtain $h_{dip} \approx -1.5h_C$. The energy barrier for the fission and fusion processes are plotted in 2.13b. The barriers are considerably lower than for the isolated stripe due to the smaller h_{eff}.

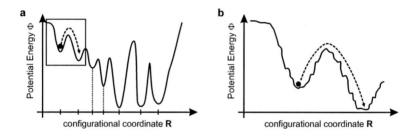

Figure 2.15: a Illustration of the potential energy landscape, **(b)** with random pinning potential.

2.7 Non-equilibrium properties

2.7.1 Glass-like and slow dynamic phenomena

In section 2.3 we have studied regular domain arrays and in section 2.5.2 constructed an equilibrium phase diagram by extrapolating the ground state properties to finite temperatures. As we have already seen in section 2.6, changing the sample's domain pattern in response to some external stimulus such as a field or temperature change, may involve the crossing of an energy barrier also in absence of domain wall pinning. The mere existence of such energy barriers raises the question under which conditions the system will be able to reach the energy minimum discussed in section 2.3. The process of equilibration may be viewed as a dissipative motion of the system in its potential energy landscape. [25, 26] In this picture, a state is associated with a local minimum of the hypersurface defined by the potential energy $\Phi(\mathbf{R})$, where $\mathbf{R}$ refers to a point in the multi-dimensional configuration space. An illustration is given in Fig. 2.15a. These minima constitute a discrete set of points $\mathbf{R}_\alpha$. For almost every point $\mathbf{R}$ we can say that it lies in basin B_α if its motion governed by the equation $\dot{\mathbf{R}} = -\nabla\Phi$ ends at $\mathbf{R}_\alpha$. At absolute zero temperature, the system relaxes to its nearest local minimum and stays there forever. At finite temperature we can describe a change of the system as a hopping from one local minimum to another one. In this process, the system has to cross an energy barrier corresponding to the lowest saddle point between adjacent basins.

At elevated temperatures, the thermal energy of the system is high enough to overcome any pass in the landscape within reasonable time and it therefore samples all states α with their Boltzmann-probability $\exp\left(\frac{\Phi_\alpha}{kT}\right)$. In such a situation the time-average $\langle O \rangle_t$ and the phase-space average $\langle O \rangle_{ps}$ of an observable O coincide, the system is ergodic. If the temperature is lowered, the system may be trapped in some region of phase space surrounded by a high enough energy barrier. Such a macro basin may still contain several basins among which the system can move. It may further contain the global energy minimum state but most probably it will not, and the system remains in the metastable state corresponding to the local energy minimum of the basin for long times t_{meta}.

In this picture the time scale at which the system is measured is crucial. During a single measurement we average over the states the system visits during the time scale t_{exp} of the experiment. If $t_{meta} \gg t_{exp}$ in the measurement, the system will appear static and we may be tempted to interpret the measured state as the equilibrium, i. e. the lowest free energy state. The characteristic time scale need not be homogeneous for the entire system, there may be different time scales associated with different physical processes. Observables related to fast processes may therefore be in equilibrium while slow observables may be at any value. In our experimental situation the magnetization M_S inside the domains is such a fast observable. Its value is connected to the spin fluctuations inside the domain and these occur on a time scale that is orders of magnitude faster than our experimental measuring time of a few milliseconds per pixel. For measurements at low enough temperature for instance the domain size L may be a slow observable since it involves changing the domain configuration, on a length scale $\approx 2L$. More complicated quantities like the structure factor of the domain pattern are most probably out of equilibrium because they rely on the global (or at least long range) arrangement of the domains.

Changing the domain configuration may require domain nucleation or collapse and certainly involves domain wall motion. The domain walls are however subject to pinning at structural defect such as steps on the substrate. [27] This (random) pinning potential Φ_p must be added to the bare potential as illustrated in Fig. 2.15b. Therefore even a region that is monotonic on the bare potential landscape may present many though shallow local minima if weak pinning is considered, leading to a microscopically stochastic, macroscopically creep-like motion of the domain walls.

In our system, global equilibration of the domain pattern entails an additional complication because the magnetostatic contribution to the total energy is *long-ranged*. In consequence, a change of the domain pattern in region A may lead to a change of the local field in region B, that adapts to the new equilibrium condition thus generating a feedback on region A and an effect on region C, etc. It has been proposed [28, 29] that long-range frustration inherently leads to self-generated glassiness, preventing the system from reaching the lowest free energy, crystalline, state.

2.7.2 Theoretical approaches to glassiness

The most salient feature of glassiness is a dramatic increase of the relaxation time as the system is cooled below the experimental glass temperature T_G. This T_G depends on the experimental procedure, e. g. on the cooling rate, and hence the glass transition is purely dynamical. Below T_G the relaxation time τ is often found to increase faster than the Arrhenius law

$$\tau = \tau_\infty \exp\left(\frac{E_B}{k_B T}\right) \tag{2.70}$$

which is the usual behaviour associated with thermal activation over an energy barrier E_B. A first empirical description of super-arrhenius behaviour goes back to the law for the temperature dependence of the viscosity η in liquids found by Vogel [30]

$$\eta = \eta_\infty^{\frac{T-T_1}{T-T_0}} \, . \tag{2.71}$$

A modified version of this law was introduced by Fulcher [31] and independently by Tammann and Hesse [32] who proposed the form

$$\eta = \eta_\infty \exp\left(\frac{E_B}{T-T_0}\right) \tag{2.72}$$

with $T_0 > 0$ being the temperature at which the extrapolated viscosity apparently diverges. This form is usually referred to as Vogel-Fulcher-(Tammann-Hesse) law. A theoretical justification of this empirical formula is however missing.

Due to the configurational disorder a liquid at a given temperature has a higher entropy than the corresponding solid. This excess entropy is labelled configurational entropy S_C. From thermodynamic measurements at $T > T_G$ the function $S_C(T)$ can be determined and its extrapolation to lower temperatures vanishes at a temperature T_K, the Kauzmann-temperature. Below T_K there is no entropic advantage of being in a liquid state with respect to the (amorphous) solid [33] . It is then tempting to identify the phenomenological T_0 with T_K and we write

$$\tau = \tau_\infty \exp\left(\frac{DT_K}{T-T_K}\right) \, . \tag{2.73}$$

For a high Kauzmann-temperature T_K the deviation from Arrhenius behaviour is strong and the parameter D is small.

On general grounds it is believed that an essential ingredient for glassiness to occur is frustration [34]. For instance, for four atoms (hard spheres) with attractive interactions in three dimensions the most stable configuration is a tetrahedron. For a larger cluster the icosahedron is found to be optimal. However, 3D space can not be tiled by regular icosahedra or tetrahedra. Therefore, for systems exceeding a certain size, the lowest energy configuration is given by a crystal. A glassy state in such a system is obtained by a quench from the melt, i. e. by cooling the liquid system so fast that it cannot crystallize. In two dimensions the situation is different. The preferred local arrangement is hexagonal and 2D space is perfectly tiled by hexagons. Since for identical particles in 2D there is no geometrical frustration, no glassiness is expected for only local interactions such as hard spheres [35] or Lennard-Jones systems [36]. Glassiness may however appear in binary mixtures [35, 36].

In systems with competing interactions on different length scales Schmalian and Wolynes [28] found a self-generated glass transition, caused solely by the frustrated nature of the interactions without the need for quenched disorder, i. e. disorder induced by random pinning centres. An exponentially large number of metastable states emerges, leading to a slow, energy landscape dominated relaxation following a Vogel-Fulcher-law (2.73).

A different connection between frustration and super-Arrhenius behaviour of the relaxation time has been proposed by Tarjus, Kivelson, Nussinov and Viot, reviewed in [37]. They start from an unfrustrated system with a critical point T^* that separates the disordered, liquid state from the ideally ordered (crystal) state. If a weak frustrating interaction is added, the critical point is avoided because frustration prohibits a long-range crystalline order, corresponding to the uniform, saturated, state in our case. However, the defects (i. e. domain walls) that break the ideal order may still order periodically, leading to a defect-ordered phase [37] as is the case for the stripe phase in the system treated here. The authors of [37] argue that

> The relevant temperature about which one can organize a scaling description of the viscous slowing down and other collective properties of supercooled liquids is that of the avoided critical point in the associated unfrustrated system. This, of course, is only meaningful if the critical point is narrowly avoided, which implies that the frustration (...) is small enough.

In the presence of domains the activation energy depends on the domain size that increases as the temperature is lowered. In a scaling analysis they propose that the relaxation time τ has an activated-like dependence on temperature, in which the energy barrier becomes temperature dependent as the system is cooled below the (avoided) critical point T^*. They propose a scaling

$$\tau = \tau_\infty \exp\left(\frac{E_\infty + \Delta E(T)}{k_B T}\right) \tag{2.74}$$

with

$$\Delta E(T) = \begin{cases} 0 & , \, T > T^* \\ B k_B T^* \left(1 - \frac{T}{T^*}\right)^\psi & , \, T < T^* \end{cases} \tag{2.75}$$

where the exponent ψ depends on the system and the parameter B is a measure of the deviation from the Arrhenius behaviour. We can rewrite this scaling for $T < T^*$ as

$$\tau = \tau_\infty \exp\left(\frac{E_\infty}{k_B T}\right) \exp\left(B \frac{T^*}{T}\left(1 - \frac{T}{T^*}\right)^\psi\right) \tag{2.76}$$

3 Experimental methods

The aim of this chapter is to give an introduction to the experimental techniques used in this work including their limitations.

3.1 Sample preparation

3.1.1 Cu-crystals

The substrates used in this study are copper single crystals[1] with the (001)-oriented surface oriented better than 0.1% and mechanically and electro-polished to a roughness depth of less than 0.03 µm resulting in an average atomic terrace width of more than 250 nm. The crystals had a square surface measuring 4 mm by 4 mm with the edges aligned either along the (100) or the (110) direction.

3.1.2 Sputtering

To remove surface contaminants and old Fe-films the crystal is cleaned by Ar^+-ion sputtering. Typically the Ar-pressure during sputtering is $2.5 \cdot 10^{-5}$ mbar and the ion energy is 1 keV. To preserve the cleanliness of the sputtering environment the turbo pumps are left running thus producing a constant flow of argon through the chamber. During this ion-milling the topmost atomic layers of the sample are gradually removed leaving behind a rough surface and a damaged crystal structure. Typically samples are sputtered during 20 minutes.

3.1.3 Annealing

Subsequent to the sputtering the sample is heated to anneal the defects introduced by the Ar-ions. The heating is done by electron bombardment. Electrons are thermo-emitted from a filament and accelerated to the sample transporter backside by a positive high voltage. The sample temperature is controlled via the heating power using a calibration curve measured with a thermocouple. Additionally, the sample temperature during annealing is monitored by pyrometry. Typical parameters are a high voltage of 500 V and an emission current of 5.5 mA leading to a heating power of 2.75 W and a sample temperature of 700 K. Typical annealing times are 30 to 60 minutes. To make the annealing process really reproducible, the emission current can

---

[1]The crystals were purchased from MaTecK GmbH, Im Langenbroich 20, D-52428 Juelich, Germany

be computer controlled by a LabVIEW[2] program. A starting time, heating ramp, annealing time and cooling ramp can be defined.

3.1.4 Cool down

After the annealing the sample cools down according to Newton's exponential law. Since the substrate temperature during evaporation influences the properties of the Fe film it is crucial to keep the experimental steps between annealing and evaporation as reproducible as possible. For this, the sample is first let to cool down on the heating stage for a fixed time, typically 45 minutes, before it is transferred to the evaporation stage. The transfer takes about 5 minutes and introduces some imprecision as it is done by hand. On the evaporation stage the sample rests for 10 more minutes to ensure a fixed delay time from the end of the annealing to the start of the evaporation.

3.1.5 Evaporation

The Fe is evaporated from a Knudsen type effusion cell [38, 39]: A tungsten crucible containing high purity Fe wire (99.99+%)[3] is heated by electron bombardment. At high enough temperature the Fe starts to sublimate and a beam of Fe atoms exits the crucible. The evaporation rate depends on the distance from the sample to the crucible and the crucible temperature. Prior to evaporation the crucible is preheated until it has reached a constant temperature. The evaporation time is controlled by a shutter that is opened and closed manually. Using a stop-watch for the timing, the accuracy of the evaporation time is ±2 seconds. A typical evaporation rate is 0.1 to 0.2 ML per minute giving typical evaporation times between 10 and 20 minutes.

Structures

By placing a mask between the evaporation source and the sample, laterally structured films can be grown [39, 40]. For this purpose, the diaphragm is placed on an inertial slider that allows a soft approach. Using this technique, films of many different shapes can be grown as long as the shape is simply connected. The most used are islands of different shapes, long strips or a half-plane produced by covering part of the substrate by an edge. The large (30 cm) sample-source distance compared to the short (10 μm) mask-sample separation gives rise to sharp edges of the structures. To ensure that diaphragm and the sample are in a relative stable position, they need to rest for a while after positioning.

---

[2]National Instruments Corp., Austin TX, USA
[3]Goodfellow Cambridge Ltd., Huntingdon PE29 6WR, UK

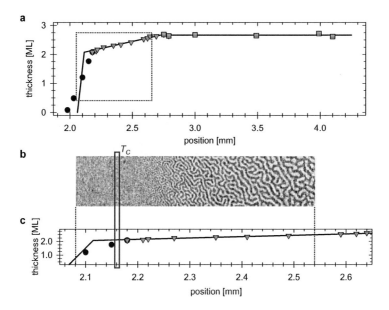

Figure 3.1: Fe wedge. **a** Thickness profile measured using AES (points) and fitted wedge shape (continuous line). **b** SEMPA image of a Fe wedge at 320 K. The red frame marks the region where the magnetic contrast disappears, i. e. where current temperature is equal to the film's Curie temperature. **c** Portion of the thickness profile marked by the dotted box in a. Correlating the position in the SEMPA image b with the sample position in the Auger profile c allows to determine the local film thickness.

Wedges

By moving the diaphragm during evaporation, wedge structures can be produced. For this purpose, the diaphragm consists of a cleaved GaAs wafer having an atomically sharp, straight edge. This diaphragm is then moved laterally to successively cover or uncover part of the sample surface. A profile of such a wedge is shown in figure 3.1. Three different regions of the wedge can be identified. An initial step, the wedge and a flat area. Note that the initial step appears broadened over 0.2 mm due to the limited lateral resolution of the Auger spectrometer. The wedge itself has a slope of about 1 ML per mm or 1:10⁷. On an atomic scale also the wedge is almost flat. The flat region of thickness 2.65 mm corresponds to the sample area that was not at all covered by the mask. In the SEMPA image the increasing film thickness correlates with an increasing domain size. We see that the variation of the domain size occurs over a length scale that is comparable to the domain size itself. In spite of the flatness of the wedge in absolute units we therefore expect that the domain pattern for a given thickness in a wedge and in a homogeneous film may differ considerably. This effect is reduced for temperatures close to the (local) Curie temperature as there the domain size becomes small.

3.1.6 Auger Electron Spectroscopy

Due to the short escape depth of the Auger electrons, Auger Electron Spectroscopy (AES) is highly surface sensitive. In this study AES is used for two purposes:
i) for checking the cleanliness of the surface between sputtering and annealing and
ii) to determine the thickness of the evaporated Fe-films. From the comparison of the intensities of the main Fe and Cu peaks in the Auger spectrum at 654 and 921 eV respectively the film thickness d can be calculated using the formula

$$d \text{ [ML]} = 4.1 \left(1 + 1.2 \frac{I_{\text{Fe654 eV}}}{I_{\text{Cu922 eV}}} \right) .$$

It can be derived in a continuum-model assuming exponential attenuation of the electron intensity with the distance travelled in the film and the substrate [6]. The parameters have been determined from calibration measurements using Scanning Tunnelling Microscopy (STM) [38].

3.1.7 Low Energy Electron Diffraction

The surface crystal structure can be determined by Low Energy Electron Diffraction (LEED). In the present study LEED is used only qualitatively to verify the crystal orientation or the surface crystallinity by observing the positions and sharpness of the diffraction spots.

3.2 SEMPA

3.2.1 Scanning Electron Microscopy with Polarization Analysis

The magnetic properties of the Fe films are measured using a Scanning Electron Microscope with Polarization Analysis (SEMPA) [41–43]. It consists of a Hitachi S-4100 SEM modified to be UHV compatible and a home built Mott-detector for polarization analysis. See figure 3.2a for a schematic drawing of the setup. The primary electron beam of the scanning electron microscope hits the sample under an angle of 45°. Inelastic scattering leads to the emission of secondary electrons. These are collected perpendicular to the incident beam by an electrostatic lens. A system of further electrostatic lenses deflects the beam by 90° in the horizontal plane. Thereby the low energetic electrons, originating from the topmost atomic layers of the sample,

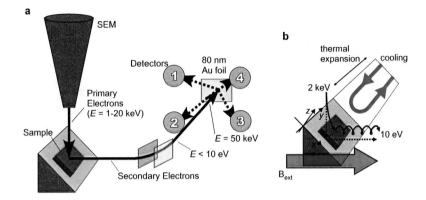

Figure 3.2: SEMPA. **a** Schematic setup. The primary electron beam from the SEM gun hits the sample under an angle of 45°. The generated secondary electrons are collected and deflected by a system of electrostatic lenses selecting the low energetic electrons. These are then accelerated to 50 keV and hit a gold foil target. The back scattering angle is spin dependent due to spin-orbit interaction. Detectors (numbered from 1 to 4, see text) count the incoming electrons. **b** Schematic drawing of the effects of the applied magnetic field and the cryostat temperature variation on the image position.

are selected. These are accelerated to 50 keV and hit a thin gold foil. The scattering angle of an incoming electron depends on its spin due to spin-orbit interaction [44,45]. Two pairs of detectors count the back scattered electrons. From the asymmetry in the number of detected electrons the spin polarization of the low energy secondary electrons can be calculated. The valence band electrons in a magnetic material are spin polarized and the electron spin is preserved if they are excited and emitted as secondary electrons [46]. Therefore the spin polarization of the collected secondary electrons is a measure for the magnetization at the actual scanning spot on the sample surface.

The out of plane (z) polarization component of the beam is measured by the detector pair 1-3 as sketched in Fig. 3.2a. It is given by

$$P_{13} = \frac{1}{S(\theta, E, d)} \frac{N_1 - N_3}{N_1 + N_3} = \frac{1}{S} A,$$

where N_1 and N_3 denote the number of electrons counted by detectors 1 and 3 respectively. $S(\theta, E, d)$ is the Sherman function [45] and depends on the scattering angle θ of the detected electrons, their energy E and the thickness of the gold foil d. It is a property of the Mott detector and is a measure for its ability to separate the two polarization directions along one axis. For instance, if $S = 0.5$, an electron beam with

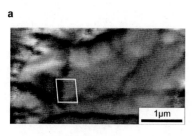
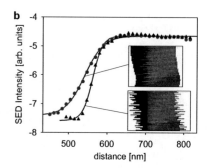

Figure 3.3: Spatial resolution of the SEMPA. **a** Topographic image recorded in SED mode at a magnification of 40k. The pixel size is 10.73 nm, each line has 400 pixels and the measurement time at each pixel is 200 μs. The white box marks the edge used to determine the resolution. **b** Line profiles across the edge marked in image a for two different alignment modes of the scan lines as shown by the insets in figure b.

100% polarization in direction of detector 1 ($P_{13} = 1$) produces an asymmetry A of 0.5. Since A is a normalized quantity, fluctuations of the incoming electron beam or effects from topography are automatically cancelled in the magnetization images, at least as long as both detectors have the same characteristics. Typical currents are 0.6 nA for the primary electron beam, 0.3 nA for the sum of all secondary electrons (1 - 10 keV) and 30 fA ($\approx 200\,000$ electrons per second) arriving on each detector. This means that only a fraction as small as $4 \cdot 10^{-4}$ of all generated secondary electrons is eventually measured in the Mott detector.

This makes clear what the main difficulty of this technique is: In order to get a statistically reasonable number of electrons on each pixel, one needs to scan quite slowly. A typical acquisition time is 10 ms per pixel. At this speed it takes about 7 minutes to record an image of 200×200 pixels. Of course, the longer the acquisition time per pixel, the better gets the signal-to-noise ratio of the picture. In change, the sample may be influenced by a higher energy dose on the small scanning spot and one has to cope with sample drift during measurement.

Obviously, $N_1 + N_3$, or $N_{Tot} = \sum_i N_i$, again gives the topographical information as in a standard SEM with secondary electron detector, but with a high surface sensitivity because only the low energy secondary electrons are measured. Analogously to the out-of-plane polarization, the detector pair 2-4 measures the in-plane component of the magnetization along the x-axis. To measure the magnetization along the other in-plane axis, the sample can be rotated by 90° around the out-of-plane axis. In parallel with the signals from the Mott-detector, the signal from the regular secondary electron detector (SED) is collected giving complementary information.

In principle the lateral resolution of the SEM is 20 nm, due to the large working distance and low acceleration voltage used. The practical resolution is determined in Fig. 3.3. Image a shows an SED image measured with 200 µs integration time per pixel and a magnification of 40k, corresponding to a pixel size of 10.73 nm. A presumably sharp edge in the image is selected (white box) and the scan lines are aligned to correct only for the contour of the edge as determined from a smoothed version of the original image (upper inset in b). For the lower inset, the scan lines have been aligned individually, in this way we eliminate the noise up to a frequency of at least 1kHz (5 pixels in the image) whereas the time difference between two subsequent scan lines is 80 ms. The data of the insets is then averaged along the vertical direction to obtain the profiles plotted in b.

The fit to the red circles corresponding to the upper inset gives an edge width of 87 ± 3 nm. The resolution determined from the profile indicated by the blue triangles is 45 ± 2 nm. Notice that for SEMPA measurements typically performed with a measurement time of 10 ms per pixel the relevant resolution is 90 nm because the time separation between neighbouring pixels is of the same scale as the time separation of different scan lines in Fig. 3.3a. The resolution is therefore limited by line frequency electromagnetic noise and mechanical vibrations of the sample stage induced mainly by the turbo molecular pumps and their cooling water supply. By switching them off the resolution can be pushed to 50 nm. In contrast, if the cryostat is in operation and the liquid He syphon is connected, additional vibrations are transmitted to the sample stage by the flow of the cooling agent and from the building floor via the dewar and the syphon.

A further issue limiting the resolution occurs when imaging larger areas: since the sample is tilted by $45°$ with respect to the horizontal, only a band of finite width in x- and extending along the y-direction can be focussed at the time.

3.2.2 Temperature control

An important feature of the SEMPA setup used is the cryostat / temperature controller. As indicated schematically in Fig. 3.2b, the sample stage is mounted on the heat exchanger of a cryostat. The heat exchanger in turn is the end of a 30 cm long rod that is fixed on an x-y-z-stage allowing the positioning of the sample with respect to the electron beam and the extraction lens of the spin analyser optics. To measure the temperature two Si-diodes are installed, one at the heat exchanger and one at the very end of the sample holder. Using liquid helium as a cooling medium the cryostat heat exchanger gets to 4.2 K easily but the temperature at the sample stage does not fall below 6 K due to radiative heating from the room-temperature environment. The temperatures of the sample itself and especially of the sample surface are more difficult to determine. A test with a superconducting V_3Si sample having a critical temperature of 16 K proved that sample temperatures below that value are reachable if care is devoted to the thermal contact between sample, sample transporter and sample stage. Using liquefied gases for cooling at temperatures considerably higher than their boiling point produces an effect in the cooling pipe that is similar to a geyser

3 Experimental methods

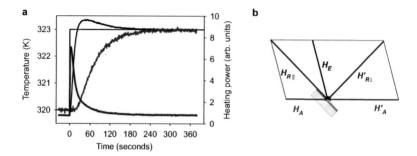

Figure 3.4: a Sample temperature (red curve), heat exchanger temperature (blue curve) and heating power (green curve, right scale) after a sudden increase of the temperature set point (black curve) at $t=0$. **b** Geometry for the magnetic field compensation. The perpendicular component of the residual earth field H_E is compensated by the applied field H_A while the in-plane component of H_E is compensated by an applied field H'_A of opposite sign. A completely field free environment can not be achieved.

eruption [47] accompanied by temperature oscillation, making temperature control very difficult. In such cases it is preferable to use cold gas as a cooling agent although the cooling power is considerably lower. With cold $N_2(g)$ a temperature of 220 K is readily reached, sufficiently low for most experiments in this work.

The fact that the exact sample temperature is not known needs to be taken into account when interpreting measured data. From a test measurement with a thermocouple attached to the sample holder (see Fig. 3.4a) we found that the sample temperature may considerably lag behind the heat exchanger. When the temperature set point is increased suddenly, the heater responds immediately and the heat exchanger reaches the new set point within a few seconds. At this time, the sample temperature has not even reacted to the change. It takes 4 to 5 minutes until the sample temperature has reached the set point value. Notice that it takes about the same time for the heater to settle at the new equilibrium value. A constant heating power is an indication that the cryostat is in thermal equilibrium.

The construction of the cryostat as a rod has important consequences for the measurement. Since the air-side end of the cryostat is attached to the UHV chamber it stays approximately at room temperature. The vacuum end is at the heat exchanger that is at variable temperature resulting in a temperature gradient along the cryostat rod. If the temperature of the heat exchanger is varied a new equilibrium gradient needs to be established along the rod which takes some time depending on the magnitude of the temperature change. Since the control of the heater is computerized, this is not a problem for maintaining the heat exchanger temperature constant. The important consequence is that the temperature change along the cryostat rod is accompanied by thermal expansion or contraction depending on the sign of the temperature change. We find that the thermal expansion coefficient for our cryostat is about 3 μm per K temperature difference at the heat exchanger. For a temperature change from room temperature to 4 K this corresponds to a contraction of the rod by 0.9 mm and a corresponding drift of the image along the y-direction. The time needed until the new equilibrium temperature distribution is reached after such a large step exceeds one hour.

This drift is very important when measuring small structures with high lateral resolution. For the measurement of extended films the consequences are less severe as it is usually not very important at which exact position the domain pattern is measured. Furthermore, since the thermal expansion coefficient is known, the thermal expansion can be compensated after every measurement by readjusting the x-y-z-stage if measuring the same sample area is important. Since the drift is along the y-direction the SEM focus is essentially unaffected.

3.2.3 Magnetic fields

Using a small Helmholtz coil located in the UHV chamber a magnetic field can be applied to the sample. As indicated in Fig. 3.2b, the magnetic field points along the axis that connects the spin analyser with the sample. In this geometry, the slow secondary electrons are forced on a helical trajectory but still reach the detector. The trajectory of the faster primary electrons is deformed to an arc of a circle whose radius is given by their cyclotron radius. As a result they hit the sample at a different position than in zero field. This deflection is along the y-axis as sketched in Fig. 3.2b and therefore the SEM focus is only affected slightly. The image shift induced by a 1 mT field is about 1.3 mm. To keep the field of view the same, the sample position can be corrected in the same way as for the temperature drift. The magnetic field also induces Larmor precession of the spin that deteriorates the spin polarization. The precession period for a 1 mT field is 35 ns while the travel time of a 10 eV-electron in the field is about 10 ns. For 1 mT the number of electrons that reach the detector is reduced considerably and SEMPA measurements as this field strength are barely possible. Since the saturation fields for the Fe/Cu(001)-system are well below 1 mT, we can use SEMPA to measure the domain pattern in an applied DC field.

Because the magnetic fields are very weak, magnetic shielding is very important. The SEMPA chamber is shielded by a μ-metal lining but even so there is a small residual earth field of about 20 μT in the chamber. By applying a small magnetic field, either the in-plane component or the out-of-plane component of the residual field can be compensated. Fig. 3.4b shows the geometric details. Because the perpendicular anisotropy of our Fe-films corresponds to a field of 100 mT [48] an in plane field on the order of 1 mT will not affect the sample. As a consequence, for the present experiment only the out of plane component of the magnetic field is relevant and the magnetic field values quoted from here onwards refer to the effective, resulting perpendicular component of the magnetic field. Its zero-value is determined directly from the analysis of SEMPA images.

Although SEMPA does not work at higher fields, the coil supports a current of 1 A leading to a field of 3 mT for long times (hours) and short pulses up to 10 A. Because the applied field deflects the primary electrons, it needs to be very stable. If the fluctuation of the beam is to be kept below 50 nm the field needs to be stable to 40 nT meaning that the current in the coil is required to be stable within 15 µA. For different types of measurements, different power supplies are used that comply with these condition. For regular DC fields we use a home built DC power supply with a range of -0.9 to 0.9 A. However due to the inductive load, fast switching of the magnetic requires high voltages. Using this power supply the minimum switching time is about 20 ms. In case faster switching was needed a commercial source measure unit[4] with a range of -100 to +100 mA and a maximum voltage of 100 V was used in combination with a digital trigger giving switching times below 1 ms.

3.2.4 AC-measurement

A special measurement mode implemented in the SEMPA-software enables time dependent measurements of the global magnetization. In this mode, the SEM beam is not controlled externally but scans the entire field of view at a rate of 50 frames per second. For a collecting time being a multiple of 20 ms, the average magnetization in the field of view is measured. In order to enhance the signal to noise ratio, the measurement can be repeated and the results from the individual loops are averaged. To increase the time resolution, also shorter collecting times down to 200 µs are possible. In that case it is important to choose a collecting time that is not a divisor of 20 ms to avoid that the acquisition is synchronized with the scanning of the beam, potentially giving rise to artefacts.

The control of the applied field is synchronized to the measurement loop and in principle the field can have any waveform as long as it can be digitized using 1024 sampling points. The simultaneous change of the magnetic field and acquisition of Mott-counts is not trivial because the magnetic field influences the sample position being measured and the number of electrons reaching the detector as discussed in the previous subsection. In practice only step changes of the magnetic field were applied because for piece-wise constant magnetic field the artefacts arising from the field change can easily be separated from the true time dependence of the magnetization.

---

[4]Keithley 236, Keithley Instruments Inc., Cleveland OH, USA

4 Equilibrium properties in zero field

The behaviour of ultrathin Fe-Films on the Cu(001)-surface without applied magnetic fields has been studied extensively by Oliver Portmann [6]. As we have seen in the theory chapter, at a given temperature T the sample can be characterized by the following parameters:

i) Film thickness d

ii) Equilibrium stripe width in zero field $L_0(T)$

iii) Saturation magnetization $M_S(T)$

4.1 General properties of ultra-thin Fe films on Cu(001)

Bulk iron at room temperature is found to be in the body-centred-cubic (bcc) phase (so-called α-iron) and exists in the face-centred-cubic (fcc) phase (γ-iron) only for temperatures above 1184 K [49,50]. In contrast, copper crystallizes at room temperature in the fcc lattice. The lattice constant of copper is $a_{Cu} = 3.61$ Å whereas for α-iron it is $a_{bccFe} = 2.87$ Å. However, the bcc unit cell only contains 2 atoms whereas the fcc unit cell contains 4. If we thus assume the density to be the same in both the fcc and bcc phase of iron, we get a lattice constant $a_{fccFe} = \sqrt[3]{2}\, a_{bccFe} = 3.62$ Å, which would match a_{Cu} quite exactly. Effectively, when Fe is deposited by molecular beam epitaxy (MBE) on the Cu(001) surface, the fcc phase can be stabilized for films not exceeding a certain thickness. Due to the slight lattice mismatch the Fe films don't grow in a perfect cubic, but rather in a slightly distorted, face-centred-tetragonal (fct) structure, which leads to an additional uniaxial anisotropy besides the surface- and interface anisotropies. As has been pointed out by Giergiel *et al.* [50] the film's morphology and its magnetic properties depend strongly on the preparation conditions. We limit the following description to films grown at room temperature as this corresponds to our experimental situation. In the (room temperature) growth of Fe on Cu(001) four morphological stages can be identified.

Investigation of films by scanning tunnelling microscopy (STM) reveal [50] that the growth mode of Fe on Cu(001) is almost perfectly layer-by-layer. When one nominal ML is deposited (i. e. one atom per surface atom), roughly 80% of the atoms are found in the first layer and none in the third. At first sight, the onset of long range ferromagnetic order is expected to be just when the first Fe-layer reaches its percolation threshold. This is, when most of the islands are connected, or - mathematically - when there is a finite probability to find an infinite path among them. This threshold in Fe on Cu(001) is estimated to be at about 0.9 ML. For example in Fe grown on W(110), 0.6 ML — all of it in the first atomic layer — lead to connected islands and are sufficient to produce magnetic order with a Curie-Temperature T_C of 190 K. [51]

In Fe on Cu(001) magnetic order is not observed until a film thickness of 1.2 ML [52] with $T_C \approx 60$ K. For this coverage, the first layer is found to be complete and the third layer still empty. [50] Probably there is some intermixing between Fe and Cu in the first layer, leading to this delayed onset of ferromagnetic order [50]. For a thickness of 1.65 ML the second layer is percolated [50] and T_C is found to reach room temperature in agreement with [6]. We find that the Curie temperature reaches a maximum for $d \approx 2.5$ ML with a value of $T_{C,\mathrm{max}} \approx 350$ K and decreases again to room temperature at $d \approx 3.5$ ML [14].

For a thickness of 4 to 5 ML the saturation magnetization abruptly decreases [53], but the magnetization direction remains out of plane. This is attributed to the fact that thicker films grow in a more perfect cubic structure, but still the surface and interface anisotropies are strong enough to keep the easy axis out of the surface plane. From 5 ML on, the film becomes unstable towards the fcc-to-bcc transition and thin needles of bcc-Fe start to nucleate. These may help to relax the rest of the film towards its fcc phase but their number is still negligible. The film can therefore still be regarded as having a fcc structure.

In the range of 10 to 12 ML, eventually the bcc phase becomes predominant, leading to a polycrystalline α-Fe film whose easy magnetization axis lies in the plane [50] as is expected from the shape anisotropy of a flat disk.

4.2 Temperature scale

4.2.1 The accessible temperature range

If we are interested in the equilibrium properties of our sample we must ensure that the sample is in thermal equilibrium, i. e. in its lowest free energy state. Reaching the new equilibrium state in response to a change of external parameters such as temperature or magnetic field may involve the crossing of energy barriers as discussed in section 2.6 . For the system to be able to relax to the new equilibrium the absolute temperature of the sample must be high enough to allow for these crossings. This sets a lower bound for the temperature range in which reliable equilibrium measurements are possible. The upper bound is obviously set by the Curie temperature T_C. As pointed out in Ref. [24] the domain pattern becomes increasingly mobile in close vicinity to T_C. Due to the limited time resolution of SEMPA this further limits the accessible temperature range.

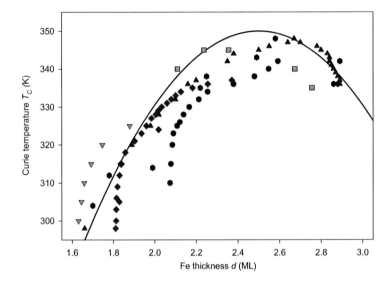

Figure 4.1: Curie temperature vs. film thickness. The symbols represent different measurements, the continuous curve is a parabola illustrating the general trend.

4.2.2 Curie temperature vs. film thickness

Within the boundary conditions just stated, it is important to tune the Curie temperature to an appropriate value. If the sample preparation and film growth conditions are kept constant, the only parameter influencing the physical properties of the film is its thickness d. This can easily be controlled by the deposition time.

The thickness dependence of the Curie temperature is summarized in Fig. 4.1. Its value for several film thicknesses may be determined simultaneously from wedge samples. However, it is not trivial to correlate the sample position in the SEMPA with the position of the Auger spectrum used for the thickness measurement. This partially explains the considerable spread of the data points for the different samples observed in Fig. 4.1. The very sensitive dependence of the Curie temperature on the exact preparation conditions further affects the quality of the measurement. We find that the Curie temperature reaches its maximum for $d \approx 2.5$ ML with a value of $T_{C,\mathrm{max}} \approx 350$ K.

If the sample temperature is kept above about 340 K for longer times, the Curie temperature is observed to rise further beyond 370 K accompanied by a strong pinning of the domains. This behaviour may be attributed to some annealing effect and is certainly to be avoided. As almost all measurements in this work were performed for d between 1.7 and 2.9 ML, the thickness dependence of T_C for $d > 3$ ML has not been investigated in greater detail.

4.2.3 Choice of the Curie temperature

The domain pattern of the sample depends on its history as is typical for systems with many metastable states. [54] In order to obtain reproducible results it is essential that this memory can be erased by putting the sample in a uniform state. As can be seen from the phase diagram in Fig. 2.11 this corresponds to either fully saturating the sample in an applied field or heating it above T_C. Although the equilibrium saturation field is rather low, the collapse field for pinned stripe domains as discussed in section 2.6.1 is high also at low temperatures. Moreover, upon reducing the magnetic field from the saturated state the system will most likely relax to some random metastable state. Given these reasons it is preferable to heat the sample above the Curie temperature where all memory is erased by thermal fluctuations. For this purpose, the Curie temperature should lie somewhere between 320 and 340 K. As can be seen from Fig. 4.1 there are two regions where this condition is met: $d \in [1.9, 2.2]$ ML and $d \in [2.8, 3.2]$ ML. Note that also samples with $d \in [2.2, 2.8]$ ML can be measured accurately as long as the sample temperature is not kept at high values for long times.

4.3 Domain size vs. temperature $L_0(T)$

As we have seen in the theory chapter, in zero magnetic field a striped domain pattern has the lowest free energy. The characteristic size of the symmetric stripe pattern is the stripe width L_0 which is the same for stripes magnetized up and down. For a film of fixed thickness d the stripe width depends only on the ratio $\delta = \frac{\sigma_w}{\lambda}$. Figure 4.2 shows typical experimental data for $L_0(T)$. The data has been recorded on a sample of thickness $d = 2.54$ ML with a T_C of 348 K. After an initial waiting time at room temperature the sample is heated step-wise and the images are recorded at constant temperature. In Fig. 4.2a the domain size L_0 in µm with its estimated error is plotted against the absolute temperature T in K. Initially a labyrinthine phase (indicated by triangles in the plot) is observed and for the first data points no change in L is appreciable within the error bars. The labyrinthine phase is discussed in more detail in section 6.1.3. At 315 K two measurements were performed giving identical images. From 317 K on, the temperature is increased in steps of 2 K and after each step 3 images of the same spot on the sample are recorded. At 323 K a transition from the labyrinth to regular stripes occurs [15], data points corresponding to stripe patterns are marked with circles. Obviously in a striped pattern the domain width is better defined than in the labyrinth leading to increased precision in determining the stripe width.

The temperature dependence of L_0 for the stripe phase is analysed in more detail in

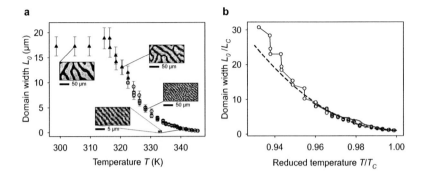

Figure 4.2: Domain width vs. Temperature, **a** experimental data in absolute units. Triangles correspond to labyrinth, circles to stripe patterns. SEMPA-Images corresponding to selected data points are displayed. **b** Experimental data for stripes compared to numerical results (red curve) and a parabolic fit (blue dashed curve).

Fig. b. For better comparison with the theory, the temperature and the stripe width are indicated in reduced units T/T_C and L/L_C where $L_C = L_0(T_C)$. The data points are connected by a line representing the sequence in which they were measured. We observe clearly that after an increase in temperature the sample initially is unchanged. During the subsequent images the domain size relaxes towards the new equilibrium value. This produces the step-like character of the curve for T/T_C between 0.93 and 0.96. While for the first two steps the 3 data points can be distinguished clearly, for $T/T_C > 0.95$ the second and third data points coincide, indicating that the system has reached thermal equilibrium. This behaviour is indicative of the decrease of the relaxation time discussed in more detail in section 6.2. For $T/T_C > 0.97$ all three data points recorded at the same temperature coincide indicating that the sample is able to adapt to the temperature change rapidly and the measurement can be considered as being reasonably isothermal. The red curve is the numerical solution of the striped mean-field model on a lattice for $\frac{J}{g} = 40$ as described in [7]. The numerical and experimental data agree reasonably well. The blue dashed curve is a parabolic fit $L_0(T) = L_C + L_1(1 - T/T_C)^2$ as is expected from the mean-field continuum approximation close to T_C [24]. The best-fit parameters are $L_C = 0.37 \pm 0.04$ µm and $L_1 = 1530 \pm 40$ µm. Also this curve reproduces the experimental results down to $T/T_C \approx 0.95$. It may therefore serve as a good approximation for the $L_0(T)$-dependence in vicinity of T_C.

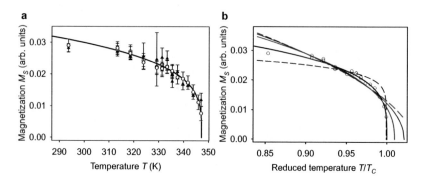

Figure 4.3: Saturation magnetization vs. temperature, **a** experimental data in absolute values. Open circles are values extracted from the histogram and black triangles are determined from digitized images. The blue curve is a best fit to the circles using a power law $M_S(T) = M_0(1-\tau)^\beta$ with $\beta = 0.26 \pm 0.015$. **b** Comparison of the fit from fig. a (blue continuous line) with numerical data (red curves) and other power laws (black curves) as discussed in the text.

4.4 Magnetization vs. temperature $M_S(T)$

The second important experimentally determined parameter is the absolute value of the magnetization inside the domain. Since at zero temperature this is equal to the saturation magnetization M_S we refer to it as $M_S(T)$. The evolution of M_S with T is shown in Fig. 4.3. The value of $M_S(T)$ may be determined in two different ways: i) by analysing the total polarization histogram and fitting two Gaussians to it (see Fig. 4.4). The separation of the centres of the two Gaussians is then equal to $2M_S$ [6]. ii) by converting the SEMPA images to digital black-and-white images and averaging the pixels in the domains identified as 'black' and 'white' separately. In Fig. 4.3a the experimental values of the asymmetry difference for both methods are plotted against the absolute temperature T. The sample had $d = 2.58$ ML and $T_C = 347.5 \pm 0.5$ K determined from the images. The temperature dependence is fitted very well by a power law

$$M_S(T) = M_0 \left(1 - \frac{T}{T_C}\right)^\beta \tag{4.1}$$

with $\beta = 0.26 \pm 0.015$ and $T_C = 347.2 \pm 0.1$ K. The T_C values as obtained from the vanishing contrast in the images and from the fit agree within their error. The exponent lies between the values expected from mean field theory $\beta_{MF} = 0.5$ and the solution of the 2D-Ising model without long-range interaction $\beta_{Ising} = 0.125$. These 3 power-laws are compared in Fig. b, the bold blue continuous curve corresponding to the fit to the data, the black dashed curve to the mean field and the black continuous

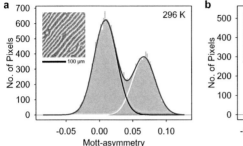 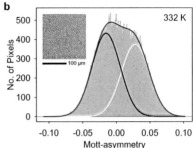

Figure 4.4: Magnetization histograms corresponding to two selected images (insets) used for figure 4.3a and fitted by two Gaussians of equal width σ. **a** Histogram at room temperature, the two peaks originating from the two magnetizations are clearly distinct. **b** At higher temperature the separation of the two Gaussians is reduced.

curve to the Ising case. The red curves represent the values as extracted from the numerical mean-field solution [7] for $\frac{J}{g} = 10$. The dashed curve represents the average of the numerical magnetization profile, the solid curve shows the behavior of the maximum of the profile. We may use (4.1) with $\beta = \frac{1}{4}$ to approximate the behaviour of M_S close to T_C.

Figure 4.4 shows the histograms for two selected images of the same sample. In figure a) a clear imbalance between black and white domains can be observed that is also reflected in the histogram. This is due to an imperfect compensation of the residual field. At higher temperature (Fig. b) the susceptibility is reduced and in consequence also the imbalance. Due to the higher temperature also the center separation of the two Gaussians used to fit the histogram has decreased which is equivalent to a lower M_S.

4.5 Magnetization profile and the domain wall

Figure 4.5a shows the magnetization profiles across one stripe domain for different temperatures. A moving average (see Fig. 4.6) has been applied to the raw data to reduce the noise. At low temperatures the profile is essentially rectangular with a constant magnetization M_S except in the narrow domain wall. As the temperature is increased, the domain size decreases and the value of M_S is reduced but the profile remains rather square up to a few K below T_C. In the vicinity of T_C the domain wall width w_{wall} becomes comparable to the domain size L_0 and as a consequence the profile acquires a rounded, cosine-like shape [7]. We can also use the value M_S as obtained from the plateau of the profile to determine the temperature dependence of

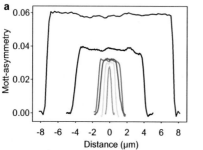 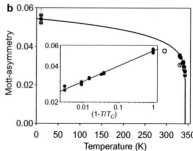

Figure 4.5: a Magnetization profiles across one domain at different temperatures. The curves correspond to 10 K (outmost), 295 K, 336 K, 338 K, 340 K and 342 K (inmost). (T_C=343 K, d=2.5 ML). **b** Magnetization vs. Temperature. The solid line is a power law fit to the blue data points. The inset shows M vs. (1-T/T_C) in a log-log plot, the straight line is the same fit as in the main graph.

M_S. Figure 4.5b shows the corresponding data points and a fit through them using a power law (4.1). The inset of fig. 4.5b displays the data for M_S against $(1 - T/T_C)$ in a log-log-plot. As can be seen, the points lie on the line that is represented by the fit with exponent $\beta = 0.129 \pm 0.004$. This agrees noticeably well with the exact critical exponent for the 2D-Ising model ($\frac{1}{8}$=0.125) but differs from the exponent 0.25 found in the last section.

To gain a quantitative insight into the temperature dependence of the domain wall width we need to establish an experimental definition of w_{wall}. For this purpose we fit the magnetization profile of the domain wall by a smooth function, a sigmoid of the form

$$M(x) = a + \frac{2M_S}{1 + \exp\left(\frac{x_0-x}{b}\right)} . \tag{4.2}$$

Here a is a constant to take into account that the Mott-asymmetry that is our measure of M is only a relative quantity, i. e. the scale may be shifted due to non-linearities of the spin analyser. The parameter x_0 gives the position of the domain wall and b is a measure of the domain wall width. To illustrate how b relates to w_{wall} we take the derivative of (4.2) (for simplicity and without loss of generality we set $x_0 = a = 0$).

$$M'(x) = \frac{2M_S}{b} \frac{1}{e^{x/b} + 2 + e^{-x/b}} \tag{4.3}$$

Obviously the derivative is a bell-shaped curve with a maximum in $x = 0$ equal to $\frac{M_S}{2b}$. Its full width at half maximum (FWHM) can be determined by setting

$$M'(x) = \frac{2M_S}{b} \frac{1}{e^{x/b} + 2 + e^{-x/b}} = \frac{M_S}{b} \frac{1}{4} \tag{4.4}$$

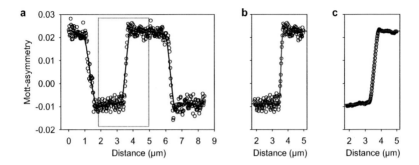

Figure 4.6: Determination of the domain wall width. **a** Magnetization profile across a sequence of domains. The black circles are the raw data points obtained by averaging 5 scan lines of 400 pixels each with a measurement time of 100 ms per pixel. The blue line is a moving average over 20 samples. The dashed frame marks the domain wall fitted in Fig. b and c. **b** Sigmoid fit of the domain wall (red line) to the raw data points giving a domain wall width $w_{\text{wall}} = 135 \pm 13$ nm. **c** Sigmoid fit to the moving average of Fig. b giving a domain wall width $w'_{\text{wall}} = 329 \pm 6$ nm.

and solving for x. We obtain that the domain wall width, i.e. the FWHM of (4.3) is

$$w_{\text{wall}} = 2\,b\,\text{Arcosh}(3) = 3.525\,b\,. \tag{4.5}$$

Figure 4.6 illustrates the procedure of fitting the domain wall to a function of the type (4.2). From the raw magnetization profile (Fig. 4.6a) the part surrounding the domain wall is selected and fitted, with the result shown in Fig. 4.6b. The fit parameters this case read

a	$2M_S$	x_0	b
-0.00877 ± 0.00013	0.0315 ± 0.00026	$3555 \pm 4\,\text{nm}$	$37.9 \pm 3.4\,\text{nm}$

giving a domain wall width $w_{\text{wall}} = 135 \pm 13$ nm. Notice that the moving average (Fig. 4.6c) introduces an additional broadening of the profile such that the domain wall width is overestimated by a about a factor 3. The pixel size p in the underlying image of Fig. 4.6 is p=21.46 nm. For the domain wall to be well resolved in the profile, the b-parameter in (4.2) should not be smaller than the pixel size, meaning that we can not resolve domain walls that are sharper than $3.525p$, in this case 75 nm. However, in section 3.2.1 we have seen that mechanical vibrations and electromagnetic noise limit the resolution of the SEM (with the imaging parameters used here) to about 90 nm. 3.2.1 The measured domain wall width is therefore rather close to the resolution limit of our setup and the obtained values should be interpreted cautiously.

Figure 4.7 shows the domain wall width as a function of temperature for two different samples with identical parameters (T_C=343 K and d=2.5 ML), one of them was measured at low temperature (red dots) and the other one (blue dots) close to T_C. The

4 Equilibrium properties in zero field

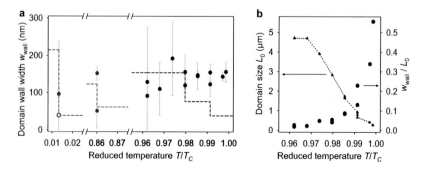

Figure 4.7: Domain wall width vs. temperature. **a** Measurements of the domain wall width $w_{wall}(T)$ at different temperatures (circles) in two different samples (blue and red, respectively), both with T_C=343 K and d=2.5 ML. the dashed lines mark the size of one image pixel. Notice the two breaks in the x-axis. **b** Experimental domain size $L_0(T)$ (triangles, left scale) vs. temperature and the ratio $w_{wall}(T)/L_0(T)$ (circles, right scale).

dashed lines indicate the resolution limit resulting from the pixel size of the images. For the high-temperature measurements no cooling was required and the turbomolecular pumps and cooling water supply were switched off to improve the microscope resolution to 50 nm, see section 3.2.1. The resolution is therefore limited by the image pixel size. For the low-temperatures liquid He-cooling is required introducing sample vibration and rising the resolution limit to ≈100 nm. In this region the resolution is limited by vibration. From Fig. 4.7a we conclude that for $T/T_C < 0.98$ we can not resolve the width of the domain wall but only state an upper bound of 100 nm that is determined by the microscope resolution. For the higher temperatures the finite width of the domain wall can be resolved. Close to T_C the domain profile has a cosine-shape and the domain wall width becomes comparable to the domain size. Fig. 4.7b shows the ratio w_{wall}/L_0 as a function of reduced temperature, the inset shows $L_0(T)$ for the particular sample measured. We observe that the ratio is small even in relatively close vicinity of T_C.

From our measurement we can not draw any conclusions about the structure of the domain wall. We do not detect any in-plane-component of the magnetization. As our time resolution is limited to a few tens of milliseconds, we can not exclude that the smooth appearance of the domain wall in our images at high temperature is due to a rapid movement of a sharp domain wall.

52

5 Static measurements in applied magnetic field

5.1 Existence of magnetic bubble domains

The patterned system responds to the applied field by expanding the majority domains at the cost of the minority domains. As we have seen in the theory section 2.3.5, at the crossing field H_{SB} a transition from stripe to bubble domains is expected. Although several experiments on Fe/Cu(001) in applied magnetic field have been performed [43, 55, 56] only recently experimental evidence for the existence of magnetic bubble domains in the 2D ferromagnetic system Fe/Ni/Cu(001) was given [57] by measuring the magnetic domain pattern in the remanent state after applying an almost in-plane field pulse. From the calculated phase diagram Fig. 2.11 we can expect to encounter the bubble phase from measurements at either constant temperature or constant magnetic field. As will become evident in this section, bubble domains are indeed the preferred domain shape at high enough applied magnetic fields.

5.1.1 Constant temperature

A series of SEMPA images illustrating the pattern transition at constant temperature is shown in Fig. 5.1. Images a to j have been recorded at high temperatures where the fluctuations are strong enough to allow for domain rearrangements. Starting from a disordered pattern of clear stripe character at (almost) zero field, image a, initially the external field results mainly in a compression of the (white) minority stripes, see image b. In this step the topology of the pattern, and therefore the total number of white domains, remains essentially unchanged although rupture or fission of selected domains has taken place and some domains have acquired a bubble-like shape. This trend is continued in image c, where the numbers of stripe-like and bubble-like domains are about equal. In image d virtually all stripe domains have broken up and the domain pattern consists of a disordered array of white bubbles on a black background. Upon increasing the magnetic field further the bubble diameter decreases but the bubble density remains approximately constant. The transition to saturation is not observed in this case because imaging is not possible in the fields required to saturate the sample at this temperature.

Images f to i show the domain transformation in decreasing magnetic field, still at high temperature. The barely visible bubble domains of image f partially merge,

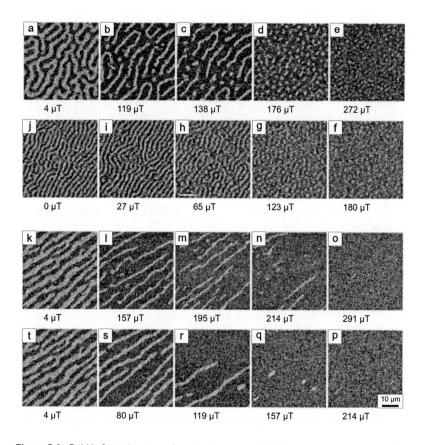

Figure 5.1: Bubble formation upon changing the magnetic field at constant temperature. The absolute value of the magnetic field for each image is indicated. Images **a-e**: increasing magnetic field at $T/T_C{=}0.994$ ($T{=}350$ K, $d{=}2.00$ ML), **f-j**: decreasing magnetic field at $T/T_C{=}0.996$ ($T{=}339$ K, $d{=}1.93$ ML), **k-o**: increasing field at $T/T_C{=}0.970$ ($T{=}317$ K, $d{=}2.03$ ML), **p-t**: decreasing magnetic field at $T/T_C{=}0.970$ ($T{=}317$ K, $d{=}2.03$ ML) Note that images p–t have been recorded before images k–o with k and t being the same image. All image sizes are 45 µm by 45 µm.

leading to more or less randomly distributed "worm"-like domains in image g. These merge further and in image h only a few bubble domains are left and the pattern clearly has acquired a stripe character. In images i and j the stripe pattern is restored. Note that the two images show the same field of view. Nevertheless, the domain patterns show considerable differences. This is an example of the high mobility and constant rearrangement of the domain pattern at this temperature [58].

At lower temperatures the situation is fundamentally different. Starting from a stripe-like pattern in essentially zero field (image k) applying the field only results in a lateral compression of the stripe domains. When comparing the magnetic field scale with the series a to e and keeping in mind that the equilibrium saturation field H_C at lower temperature should be lower, we see that image l should already show a bubble pattern if the system was in equilibrium. Images m and n show that the transition to the saturated state (o) does not proceed via the bubble phase but by collapsing of the stripe domains [59]. Taking into account that this series has been recorded only 9 K below T_C it is not surprising that the bubble pattern has not been observed in previous experiments.

When reducing the magnetic field from the saturated state (p) at low temperature only a small number of domains nucleate, see image q. Note that image q has been recorded at the same magnetic field as image l. From comparison of the two it is evident that at least one of them corresponds to a non-equilibrium state. The initial seed domains expand along one direction (r) and upon lowering the field further, more stripe domains enter the field of view (s) until the weight of up and down magnetized domains is about equilibrated in image t.

5.1.2 Constant magnetic field

The pattern evolution while changing the temperature in constant field is investigated in Fig. 5.2. The sample is scanned while the temperature is varied at a fixed rate. With a cooling/heating rate of 0.5 K/min and acquisition of each image taking 400 seconds, each one of the images a to p comprises a temperature variation of 3.2 K. A different position in the image is therefore equivalent to a different sample temperature. The temperature scale is indicated in terms of the reduced temperature in Fig. 5.2 and the images are placed accordingly. Note that images q to u have been acquired at a slightly lower heating rate of 0.4 K/min and have been scaled by a factor of 0.8 to fit on the same temperature scale. For high magnetic fields (241 μT, images a-e), no stripes are observed. When cooling from above T_C (a), bubble domains appear in image b whose density monotonously decreases with decreasing temperature until a homogeneous, saturated state is reached (c). For this field, the process is completely reversible, as can be seen in images d and e recorded upon heating.

For intermediate values of the magnetic field, (images f-h) the sequence of phases upon cooling is paramagnetic-stripes-bubbles-saturated. Clearly, the transition from bubbles to the uniform state occurs at lower temperature than at higher fields. Upon heating a different behaviour is observed. The sample remains saturated until a tem-

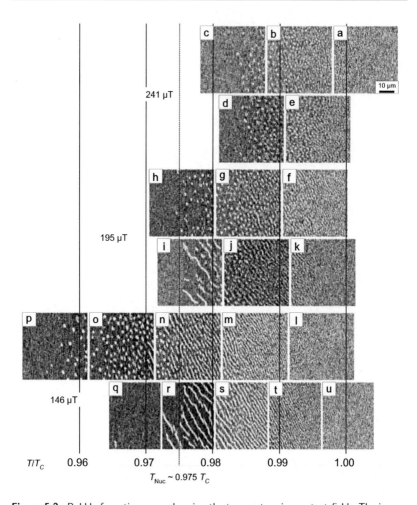

Figure 5.2: Bubble formation upon changing the temperature in constant field. The images a-p were acquired at a constant cooling/heating rate of 0.5 K/minute, leading to a continuous temperature variation within each image. Image **a-c**: cooling in 241 µT, **d,e**: heating in 241 µT, **f-h**: cooling in 195 µT, **i-k**: heating in 195 µT, **l-p**: cooling in 146 µT, **q-u**: heating in 146 µT at a rate of 0.4 K/min. The images q-u are scaled by a factor of 0.8 along the horizontal direction to fit on the same temperature scale. The reduced temperature scale is indicated. All image sizes are 45 µm by 45 µm. The film thickness is d=2.22 ML, all images were measured on the same sample.

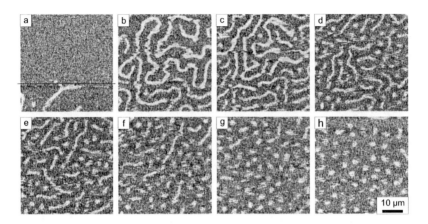

Figure 5.3: Relaxation of an irregular, string-like domain pattern to a bubble pattern in a constant magnetic field of 156 μT. Each image is scanned from top to bottom. During the acquisition of image **a** the sample is heated from 332.5 K to 335.5 K. In Image **b** the temperature settles to 336.5 K and is then held constant for the remaining images. All image sizes are 45 μm by 45 μm, the film thickness is d=2.00 ML.

perature T_{Nuc} is reached and suddenly string-like domains appear on the entire sample (image i). Upon further heating, these domains decay into bubbles (Fig. j) which at higher temperature transform to stripes that in turn disappear as T_C is reached (Fig. k). The transition from saturation to bubbles via stripe-like domains is discussed in more detail in the next subsection.

At lower fields, the same phase sequence is observed upon cooling. Because the temperature interval for each phase is wider and the domain size at lower temperature is larger, the phases can be identified more clearly (images l-p). Upon heating, again the sample remains saturated up to T_{Nuc} which in this case lies above the bubble-stripe transition temperature (compare images n and r). Therefore the nucleated string-like pattern is stable and the stripe phase persists upon further heating up to T_C.

5.2 Details of the pattern transformations

5.2.1 From saturation to bubbles upon heating in constant field

At high applied fields and rising temperature the transition from saturation to the bubble pattern proceeds by direct nucleation of circular domains. At low fields the bubble phase is avoided completely and the nucleation directly leads to a striped pattern. The domain nucleation process at intermediate fields is presented in detail

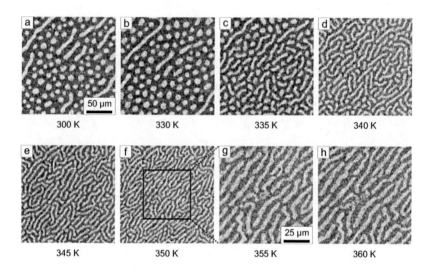

Figure 5.4: The sequence of SEMPA images shows the domain pattern transformation on heating a bubble domain pattern from 300 K up to 360 K. The magnetic field is 22 µT. Image sizes are 182 µm by 182 µm for **a** to **f** and 91 µm by 91 µm for **g** and **h**. Film thickness is $d=2.16$ ML.

in figure 5.3. An initially saturated sample is heated in a constant magnetic field of 156 µT with a constant rate of +1 K per minute. When the nucleation temperature marked by the red dotted line in image a is crossed, suddenly reversed white domains appear on the sample. At this point the temperature is halted and kept constant during the remaining images. As can be seen from the contiguous domain marked yellow in image b, these initial domains rapidly extend over large areas in agreement with the observations by Cape and Lehman [20] in thick garnet films. In our case thermal fluctuations are strong enough to break up the string-like domains and increase the total number of domains. This process can be observed already in image c where the many coloured domain segments originate from the single domain marked in b and is more pronounced in images d, e and f. Finally, in g and h only bubble domains of approximately circular shape are left. When comparing the last two images carefully, we observe that the bubbles are not at the same position indicating that the bubbles form a (viscous) liquid rather than an amorphous solid or even a crystal.

5.2.2 Bubbles to stripes upon heating in constant field

The transformation from bubbles to stripes in a constant field of 22 µT is investigated in Fig. 5.4. The sample had $d=2.16$ ML and a Curie temperature above 350 K. To

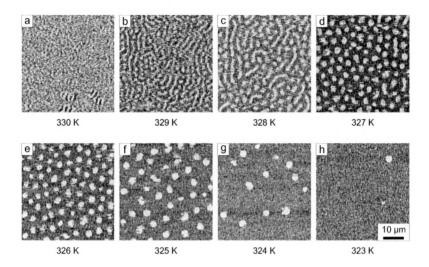

Figure 5.5: The sequence of SEMPA images shows the domain pattern transformation on cooling from above the Curie temperature T_C to saturation. The magnetic field is 95 µT and all image sizes are 45 µm by 45 µm. The film thickness is d=1.91 ML.

aid following the pattern transformation, the same domain is marked yellow in all images. Image a and b show that at low temperature the domain pattern is essentially unaffected by the temperature change of 30 K, except for the shift due to the thermal expansion of the sample holder, as discussed in section 3.2.2. They both show a pattern of essentially randomly arranged round domains. In b the size distribution is narrower and the domains have a more circular shape. From image b to c the bubble domains elongate and no preferential direction can be observed for this elongation. In figure d the aspect ratio of the domains increases further and in image e the domains start to preferentially align along the same direction. Note that this alignment is achieved by the growth in length and shrinking in width of the domains only. The number of domains remains constant and no domain merger or fission occurs. In figure f the red box marks the field of view of images g and h. The domain pattern in the last two images is again very similar. Notice the high sample temperature of 360 K for image h. At these temperatures the annealing effect described in section 4.2 plays an important role, thus questioning the reliability of the corresponding measurements.

5.2.3 Stripes to bubbles to saturation on cooling in constant field

A detailed measurement of the domain pattern variation upon cooling from above T_C in a constant applied field is shown in Fig. 5.5. The sample had d=1.91 ML and T_C=331 K. The sample was cooled in discrete steps of -1 K and after the temperature was allowed to settle during 3 minutes the acquisition of the next image was started. All images show the same field of view. The image **a** taken at 330 K shows small regions of stripe domains marked in yellow and blue. In image **b** a mostly striped pattern can be observed, intermixed with some bubble domains. The next image, **c**, shows a mixture of elongated domains and approximately circular bubbles with a larger average domain size than in **b**. At a temperature of 327 K, in image **d**, except for a few all domains have a circular shape. In figure **e** only circular domains are left. As is obvious from the image, the bubbles are not arranged in a hexagonal lattice as would be expected from the equilibrium considerations (section 2.3.5) but they are rather disordered, see section 5.3.5 for more details. Since the fast scanning direction is from left to right, the fact that domains move during scanning manifests as horizontal streaks in the image. Two such domains are marked yellow in **e**. This mobility is also observed in image **f**, for example in the yellow coloured domains. From **e** to **f** and **g** the bubble density decreases with decreasing temperature. In image **h** only two reversed domains are left and one image later the entire field of view was homogeneously magnetized.

5.3 Quantitative analysis

5.3.1 Geometric magnetization vs. temperature

As we have seen in the theory section in agreement with reference [21], the equilibrium domain pattern is closely linked to the geometric magnetization $\overline{m}$ as defined in equation (2.14). By determining $\overline{m}$ from each SEMPA image and combining many measurements similar to those of Fig. 5.2, a more quantitative insight into the pattern formation can be gained. The result is shown in figure 5.6 where the average $\overline{m}$ is plotted against the average reduced temperature T/T_C of each image. Within each series the applied magnetic field is held constant and the temperature is decreased at a fixed rate. The $\overline{m}_H(T)$-curves are labelled with their corresponding resulting field value $\mu_0 H$ in μT. From the set of curves the very high sensitivity to magnetic fields is evident. Even for a magnetic field as low as 4 μT ($\hat{=}$40 mGauss) a significant deviation from the $\overline{m}$=0-line can be observed.

Since the domain pattern itself is a very sensitive probe for the magnetic field it was used to determine the compensation field needed to produce a situation with zero net-perpendicular field. The situation of zero field corresponds to an applied compensation field of 11.5 μT. Without compensation the sample is therefore subject to a residual perpendicular field of -11.5 μT leading to a considerable excess of white magnetized areas at lower temperatures as has been observed earlier [15] and will be discussed

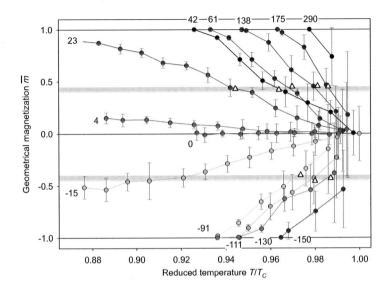

Figure 5.6: Geometrical magnetization $\overline{m}$ as a function of temperature measured upon cooling in a constant applied magnetic field. The triangles mark the experimentally observed transition from a striped to a bubble pattern, the grey bars indicate the region where the cross-over is theoretically expected. The applied field in µT is indicated for every curve.

later, c. f. section 6.1.2.

For intermediate applied fields ($30 \, \mu\text{T} < |\mu_0 H| < 200 \, \mu\text{T}$) a transition from stripes to bubbles is observed at $|\overline{m}| = 0.436 \pm 0.022$ in excellent agreement with the $\overline{m}_{cross} = 0.44$ predicted by Ng and Vanderbilt [21, 60] as discussed in the theory section 2.3.5. As can be seen from Fig. 5.8a, the magnetization at the transition depends weakly on the value of the constant applied field, with the random scattering larger than the systematic deviation. For lower fields the transition to a bubble-like pattern tends to take place at lower $\overline{m}$. This can be attributed to the lower temperature at which the transition happens. With decreasing temperature, the equilibrium domain pattern periodicity and domain size increase. Under the constraint of a constant total number of domains the system can adapt to the new equilibrium spacing by contracting the stripes along their length while expansion along the transverse direction leads to larger domains. Both effects result in a more bubble-like appearance of the domain pattern. This favours the bubble state with respect to the true equilibrium conditions.

At high applied fields the stripe phase is avoided completely as pointed out in the context of Fig. 5.2. At very low fields instead, the transition from to the bubble phase to saturation (see the 23 µT curve) or even the transition from stripes to bubbles

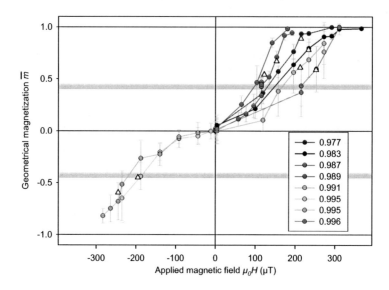

Figure 5.7: Geometrical magnetization $\overline{m}$ as a function of the applied magnetic field at constant temperature. The triangles mark the experimentally observed transition from a striped to a bubble pattern, the grey bars indicate the region where the cross-over is theoretically expected. The reduced temperature for every curve is indicated in the legend.

(-15 µT) is kinetically avoided. For a reduced temperature below about 0.9 the sample remains in its state for long times and on the experimental time scale of minutes to a few hours the domain pattern appears frozen, see section 6.1.1

5.3.2 Geometric magnetization vs. applied field

Similar to the previous subsection, we can investigate the geometric magnetization as a function of applied field at constant temperature. Since in this case the equilibrium domain size does not change, a change in the topology of the pattern involving domain fission or fusion is necessary for the sample to adapt to the new equilibrium state. As pointed out before, this is only possible at elevated temperatures. In Fig. 5.7 a set of $\overline{m}_T(H)$-curves is shown, two for negative applied field and 6 for positive. The legend specifies the reduced temperature T/T_C for each curve. In contrast to the constant field case discussed before, the sequence of the curves is not monotonic. The saturation field depends on the film thickness, i. e. T_C, and also on the absolute temperature for dynamic reasons. The transition from stripes to bubbles occurs at considerably higher fields and $\overline{m}$ than theoretically expected, on average at $\overline{m} = 0.55 \pm 0.10$.

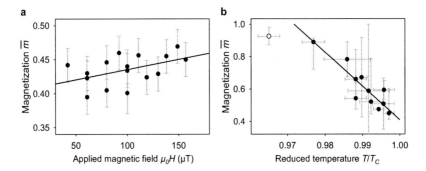

Figure 5.8: Geometrical magnetization at the cross-over from stripes to bubbles $\overline{m}_{SB}$ as determined upon cooling at constant applied field (**a**) and by increasing the magnetic field at constant temperature (**b**).

However, the magnetization at the transition is not randomly scattered about the average but presents a clear systematic dependence on temperature, see Fig. 5.8b. In contrast to the constant-field case, here the equilibrium domain spacing changes only weakly. The main process leading from stripes to bubbles is breaking up of the stripes. At high temperatures this readily happens, and $\overline{m}$ is found to be close to the theoretical value. With decreasing temperature, the crossing of the energy barrier involved in the domain fission becomes increasingly difficult, leading to a cross-over to bubbles at higher fields and higher $\overline{m}$. For the data point at lowest T/T_C (open circle in Fig. 5.8b) the transition happens exclusively by longitudinal contraction, see Fig. 5.2k-o. The $\overline{m}_{SB}$ vs. T/T_C behaviour of the other data points can be putatively fitted by a straight line as indicated in the plot. The fit extrapolates to $\overline{m}_{SB} = 1$ at $T/T_C \approx 0.972$ and to $\overline{m}_{SB} \approx 0.41$ at $T = T_C$, both with considerable uncertainty. This suggests that for $T/T_C < 0.972$ the stripes to bubbles transition by domain fission is suppressed. As pointed out in the context of Fig. 5.1, domain nucleation is not observed below $T/T_V \approx 0.975$ which is in reasonable agreement with the limiting temperature found here. This reflects the fact that the energy barriers involved in domain nucleation, fission and fusion are similar, in agreement with the estimates in section 2.6.

5.3.3 Collapse plot

In the theory we have seen (section 2.5.1) that the geometric magnetization follows a general scaling law,

$$\overline{m} \propto H \cdot L_0 \cdot M_S^{-1}. \tag{5.1}$$

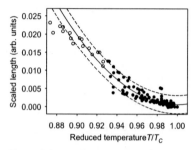

Figure 5.9: Fit of the scaled domain period vs. temperature as discussed in the context of eq. (5.5)

Eq.	a	b
(5.5)	1540 ± 260	2950 ± 500
(5.6)	1470 ± 30	
(5.7)	1380 ± 70	2800 ± 250
(5.8)	1340 ± 100	3300 ± 600

Table 5.1: Best-fit parameters for the scaling law (5.3) as obtained from the different procedures indicated by the equation number.

We trivially know the applied field $\mu_0 H$ for each measurement and from the sample temperature T and the sample's Curie temperature T_C we can readily determine the reduced temperature T/T_C. From each SEMPA image we can extract $\overline{m}$ by converting it to a digital, black and white image as in Fig. 5.12. Further we have found in chapter 4 that we may use the expressions

$$L_0(T) = L_C + L_1(1 - T/T_C)^2 \qquad \text{and} \qquad M_S(T) = M_0(1 - T/T_C)^{\frac{1}{4}} \qquad (5.2)$$

to approximate $L_0(T)$ and $M_S(T)$. We therefore make the following ansatz for the scaling of $\overline{m}$

$$\overline{m} = \frac{1}{a} \frac{\mu_0 H \left(1 + b(1 - T/T_C)^2\right)}{(1 - T/T_C)^{\frac{1}{4}}}, \qquad (5.3)$$

where

$$a = \eta \frac{M_0}{L_C} \qquad \text{and} \qquad b = \frac{L_1}{L_C}. \qquad (5.4)$$

Here η is an irrelevant proportionality constant relating the applied magnetic field $\mu_0 H$ in µT to the magnetization M_S in arbitrary laboratory units. The reason for this choice of a will become evident in the next subsection. In order to obtain the parameters a and b we represent the data in a different form, totally equivalent to (5.3)

$$\frac{\overline{m}(1 - T/T_C)^{\frac{1}{4}}}{\mu_0 H} = \frac{1}{a} \left(1 + b(1 - T/T_C)^2\right). \qquad (5.5)$$

Here the left-hand side corresponds to a scaled domain period and is proportional to $L_0(T)$. The data is presented in this way in Fig. 5.9. The open symbols at low

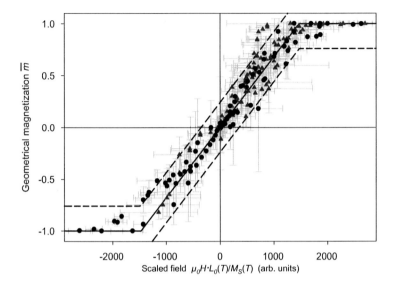

Figure 5.10: Collapse plot illustrating the scaling law found for the geometrical magnetization $\overline{m} \propto H\,L_0\,M_S^{-1}$. Blue circles correspond to data points acquired by cooling in constant field, red triangles to measurements at constant temperature.

temperature have been excluded from the fit as there the domain size is likely to be out of equilibrium, see section 6.1.1, and the parabolic approximation is expected to be valid only close to T_C.

From the best fit we obtain the parameters listed in the first row of table 5.1.

Figure 5.10 shows a collapse plot using (5.3) with the value for b obtained from the fit in Fig. 5.9. The equation used for the fit takes into account that $|\overline{m}|$ can not become larger than one. It is

$$\overline{m} = \operatorname{sgn}(\overline{H}) \min\left(1, \frac{1}{a}\,|\overline{H}|\right) \qquad \text{with} \qquad \overline{H} = \frac{\mu_0 H\,(1 + b(1 - T/T_C)^2)}{(1 - T/T_C)^{\frac{1}{4}}}\,, \quad (5.6)$$

being the scaled applied field. The geometric magnetization $\overline{m}$ is plotted against the scaled applied field. The straight line shows the best fit and the dashed lines indicate its 95% prediction band. The vertical errors are due to the uncertainty in determining $\overline{m}$ from the images whereas the horizontal errors result mainly from the uncertainty in determining T_C that propagates to M_S and L_0. At higher temperatures the measurement is less reliable for two main reasons. Due to the vanishing M_S and increasing domain mobility it is more difficult to image the domain pattern resulting in a larger $\Delta \overline{m}$. Additionally we have that $\Delta M_S \to \infty$ as $T \to T_C$ resulting in a large uncertainty along the horizontal axis. Nevertheless we observe that the proportionality constants a obtained from (5.5) and (5.6) agree reasonably well (see table 5.1) and that the data points in Fig. 5.10 fall on the straight line within their (considerable) experimental uncertainty.

5.3.4 Phase diagram

From a SEMPA image series recorded in changing temperature at fixed field $\mu_0 H$, see Fig. 5.2, we can extract the (reduced) transition temperatures from stripes to bubbles (T_{SB}) and from bubbles to saturation (T_{Sat}). Similarly we can extract the transition fields h_{SB} and h_{Sat} from series measured at constant temperature with changing field (Fig. 5.1). Determining the transition point from the SEMPA images by visual inspection obviously introduces some arbitrariness in the values. This is taken into account by stating an upper and lower bound for the transition temperatures and -fields. These bounds are represented in Fig. 5.11 by the error bars. An additional possible source of error is the uncertainty in the determination of the Curie temperature T_C. We find the saturation temperature or -field as the point of transition to the uniform, saturated state. By combining the results from several series we obtain the transition lines from stripes to bubbles and from bubbles to saturation and can construct the pattern phase diagram in T-H-space, see Fig. 5.11. The blue continuous line is a fit to the saturation transiton line following the scaling law (5.3). As for saturation $|\overline{m}|=1$, the expression reduces to the form

$$H_{\mathrm{Sat}} = a_{\mathrm{Sat}} \frac{(1 - T/T_C)^{\frac{1}{4}}}{1 + b(1 - T/T_C)^2} \, . \tag{5.7}$$

The best-fit parameters for a and b are listed in table 5.1. For the transition from stripes to bubbles we have the expression

$$H_{\mathrm{SB}} = a'_{\mathrm{SB}} \frac{(1 - T/T_C)^{\frac{1}{4}}}{1 + b(1 - T/T_C)^2} \qquad \text{with} \qquad a'_{SB} = a\,\overline{m}_{\mathrm{SB}}. \tag{5.8}$$

Notice that from fitting the above form we can determine only the product $a \cdot \overline{m}_{\mathrm{SB}}$. From the fit we find $a'_{\mathrm{SB}} = 588 \pm 42$. Due to the good agreement between the experimental and theoretical value for $\overline{m}_{\mathrm{SB}}$ found in section 2.3.5 we may use the theoretical value to calculate $a = a'/0.44$ as given in table 5.1. Additionally we may extract the cross-over magnetic field from the ratio $\frac{H_{\mathrm{SB}}}{H_{\mathrm{Sat}}}$. Because b should be independent of the transition line we look at, we can determine the ratio of the transition fields simply as

$$\frac{h_{\mathrm{SB}}}{h_{\mathrm{Sat}}} = \frac{a'_{\mathrm{SB}}}{a_{\mathrm{Sat}}} = 0.425 \pm 0.037 \, . \tag{5.9}$$

For comparison with the theory we have to keep in mind that the field at which saturation experimentally occurs is not given by the equilibrium saturation field h_C

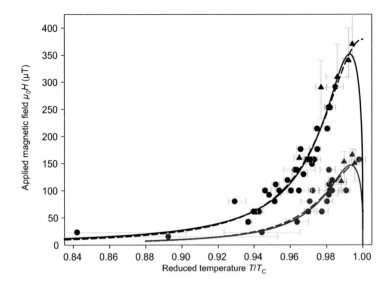

Figure 5.11: Pattern-phasediagram for ultrathin Fe-films on C(001) in T-H-space. The transition from stripes to bubbles is shown in red, the transition from bubbles to saturation in blue. Circles correspond to measurements taken in constant applied field, triangles are measurements taken at constant temperature.

but by the bubble-collapse field $h_{\text{collapse,B}}$. We find therefore theoretically, see (2.47) and (2.56),

$$\frac{h_{\text{SB}}}{h_{\text{Sat}}} = \frac{0.545\, h_C}{\frac{e}{2} h_C} = 0.401 \tag{5.10}$$

which compares favourably with our experimental results.

The dashed lines in Fig. 5.11 are fits to the data neglecting the temperature dependence of M_S i.e.

$$H_{transition} = a \frac{1}{1 + b(1 - T/T_C)^2} \tag{5.11}$$

giving the parameters for saturation $a_{\text{Sat}} = 378 \pm 16$ and $b_{\text{Sat}} = 1420 \pm 140$, Stripes-Bubbles: $a_{\text{SB}} = 155 \pm 9$ and $b_{\text{SB}} = 1520 \pm 320$. From this we can again calculate the ratio of the cross-over and saturation fields giving

$$\frac{h_{\text{SB}}}{h_{\text{Sat}}} = \frac{a_{\text{SB}}}{a_{\text{Sat}}} = 0.410 \pm 0.029 \tag{5.12}$$

which is confirms the very good agreement with the theoretical prediction. The data is not reliable enough to discriminate between the two laws (5.8) and (5.11). In consequence, the down-bend of the transition lines close to T_C must remain speculative at this point. Furthermore, it falls within the temperature region of mobile domains [24] and can therefore not be observed directly using our technique.

Nevertheless do our results establish unquestionably that the phase boundaries curve *upwards* over most of the temperature range. This behaviour is qualitatively different from the phase diagram proposed by Garel and Donaich [8] for thicker films and widely assumed to be general [16]. The positive slope of the transition lines is a consequence of the L_0^{-1}-scaling of the critical fields h_{SB}, h_C and $h_{\text{collapse,B}}$ present in ultra-thin, 2-dimensional films with $w, L \gg d$ as pointed out in section 2.5. This scaling outweighs the effect of the decreasing M_S which tends to reduce the field scale. For thicker films with $w, L \ll d$ the critical fields scale with $4\pi M_S$ and are essentially independent of the domain size [12, 19, 20]. In consequence the $M_S(T)$-behaviour dictates the transition line, leading to *down*-bent phase boundaries as predicted in [8].

A phase boundary with positive slope in qualitative agreement with our findings has been predicted by Abanov *et al.* [23] for the 2-dimensional case based on a model involving elastic constants. That calculation did however only consider stripe domain patterns and therefore missed the bubble phase.

The bubble phase has been predicted also in two dimensions by Ng and Vanderbilt [21, 60] and has been found in Monte Carlo simulations on a triangular lattice [61]. In simulations on a square lattice the appearance of the bubble domain state seems to be hindered by the strong four-fold anisotropy induced by the underlying lattice. Notice that in our samples even a single small domain with a diameter of 300 nm contains on the order of 10^6 spins and it is therefore not surprising that the effect of the fcc-square lattice is orders of magnitude weaker than in any feasible simulated system. In consequence the more isotropic triangular lattice may indeed represent a more realistic approximation to the experimental situation.

5.3.5 Characterizing the bubble phase

The bubble phase is analysed in more detail in figure 5.12. Image a shows the same data as Fig. 5.5e. Image b shows the digitized, black-and-white version of it. From this image the magnetization $\overline{m} = 0.49$ can be determined easily by counting black and white pixels. Image c shows the self-correlation

$$g(r_j) = \sum_i m(r_i + r_j)m(r_i) \tag{5.13}$$

where i and j are two-dimensional indices running over all pixels of the image b, hence $m(r_{ij}) \in \{-1, 1\}$. In the central area of the correlation image, the circular bright spot is representative of the average bubble and is surrounded by a dark ring and again a bright ring. The bright ring shows the typical bubble to bubble distance and at first sight no angular anisotropy is observed, indicating the absence of order in the bubble

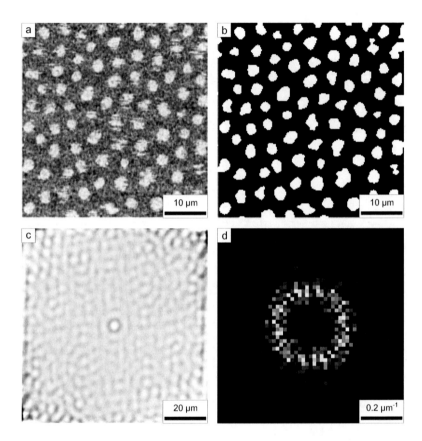

Figure 5.12: (Dis-)order in the bubble phase. **a** SEMPA image of bubble domains measured at a constant temperature of 326 K (T/T_C=0.985). **b** Black-and-white version of image a obtained by first smoothing the original data using a moving average filter and thresholding the smoothed image. **c** pair correlation function $g(\vec{r})$ as defined in (5.13) calculated from image b. The contrast enhancement at the borders corresponding to large distances $|\vec{r}|$ is an effect of the finite size of the analysed image. **d** Part of the discrete fast Fourier transform (FFT) of image b showing the central region around $\vec{k}$=0.

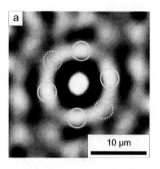
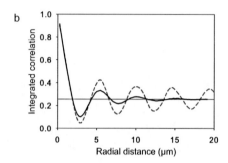

Figure 5.13: Correlation in the bubble phase. Image **a** shows the pair-correlation function $g(\vec{r})$ around $\vec{r} = 0$. A typical bubble diameter of 3 µm can be identified from the central bright spot while a typical nearest neighbour distance of 5 µm can be inferred from the radius of the first bright ring. A weak hexagonal order is observed as indicated by the yellow circles and discussed in the text. Figure **b** shows the radial, i.e. angle-integrated, pair correlation function $g(|\vec{r}|)$ as calculated from the correlation function 5.12c (blue solid line) and compared to the radial pair correlation function expected from a perfect hexagonal bubble lattice with identical domain size and nearest neighbour distance (red dashed line).

arrangement. Image d shows the Fourier transform of image b. The Fourier spectrum is characterized by a ring of radius 0.2 µm^{-1} and is isotropic in angle within its noise level.

The correlation function is analysed further in figure 5.13. Image a shows a zoom-in to the central region of Fig. 5.12c with enhanced contrast. A close look at the inner bright ring reveals four bright features marked by yellow circles indicating a weak hexagonal order of the bubbles. The other two spots originating from the hexagonal structure should appear at the positions marked with dotted circles. They can not be observed in the image. Figure b shows the angle-integrated correlation function $g(r)$. The oscillation in the correlation can be observed up to the third nearest neighbour. For longer distances, the correlation reaches the constant value $g = \overline{m}^2 \approx 0.25$ expected for randomly distributed domains.

More insight into the arrangement of the domains can be obtained from the Voronoi construction [62]. We find the center of each domain by averaging the coordinates of its pixels. The Voronoi polygon associated with a given domain i having its center at $\vec{r}_i$ may be constructed as the smallest polygon containing the point $\vec{r}_i$ that is formed by the perpendicular bisectors of the connecting lines $(\vec{r}_j - \vec{r}_i)$ to the surrounding points j at positions $\vec{r}_j$[1]. Note that for a given set of points their Voronoi construction is unique. For each cell the number of near neighbours is immediately given by its

---

[1] We generate the Voronoi construction using the 'voronoi'-function coming with MATLAB Release R2007b by The MathWorks Inc.

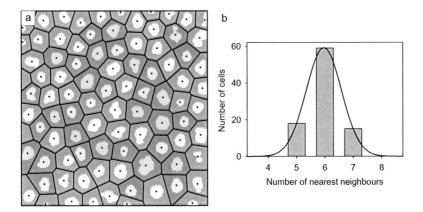

Figure 5.14: Analysis of local order in the bubble phase. **a** shows the Voronoi construction associated with image 5.12b. The black dots mark the center of mass for each domain, the black lines show the borders of the Voronoi cells. Cells with 5 or 7 nearest neighbours are shaded yellow or blue respectively. The plot **b** shows a histogram of the number of nearest neighbours.

number of edges. The construction is shown in Fig. 5.14a where the colour coding of the cells corresponds to the number of nearest neighbours. Cells with the expected 6 nearest neighbours (n.n.) are white, cells with 5 n.n. are yellow and cells with 7 n.n. are blue. The histogram in figure b shows the distribution of the number of nearest neighbours. The fit is a Gaussian centred at 6.0 with a standard deviation of 0.6. The spread in the number of nearest neighbours is a clear indication of disorder. [34] As becomes evident from Fig. 5.14a, the 5- and 7-neighbour cells mostly appear in conjugated pairs.

A further way to classify the (dis-)order in the bubble arrangement is the bond-orientation correlation. By connecting the neighbouring domain centres we obtain the Delaunay construction which is dual to the Voronoi construction introduced before. The connecting lines represent the bonds between the domains. In a perfect lattice, each domain is connected to 6 neighbours by bonds at respective angles of 60°. Measured against some common axis, e. g. the x-axis, in the ideal case the bond angle ϑ_j to the j'th neighbour can therefore be represented as $\vartheta_j = j\frac{2\pi}{6} + \varphi_0$. Here φ_0 is the global angle by which the triangular lattice is rotated w. r. t. the x-axis. To find deviations from this perfect arrangement we may associate with each domain at position $\vec{r}_i$ its local bond-orientation parameter [34, 36]

$$\psi_6(\vec{r}_i) = \frac{1}{z_j} \sum_j^{z_j} e^{i\,6\,\vartheta_j} , \qquad (5.14)$$

71

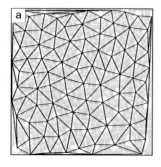 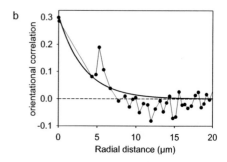

Figure 5.15: Bond-orientational order in the bubble phase. **a** Delaunay triangulation (black lines) connecting each bubble centre to its neighbours, as used for determining the orientational correlation. **b** Bond-orientational correlation (connected dots) as defined in (5.15) vs. radial distance. The orientational order decreases on the same length scale as the pair correlation. The solid line is an exponential fit to the data.

where the sum runs over all z_j nearest neighbours of domain i and the ϑ_j are the measured bond angles. To find out more about the range of the orientational order we introduce the bond orientation correlation function

$$g_6(\vec{r}) = \langle \psi_6^*(\vec{r}_i)\psi_6(\vec{r}_i + \vec{r})\rangle \tag{5.15}$$

where ψ^* denotes the complex conjugate of ψ and $\langle...\rangle$ the average over the system. Due to the complex conjugate the common angle φ_0, drops out and we are left with the relative deviations from the ideal angles. As we have seen before, in our case the bubble positions are fairly disordered and we expect therefore the correlation function $g_6(\vec{r})$ to depend only on the distance $r = |\vec{r}|$. Another consequence of the positional disorder is that the bubble-to-bubble distance is different for every pair of domains and the averaging over identical distances is not possible. Instead, we calculate the product $g_6(r_{ij}) = \psi_6^*(\vec{r}_i)\psi_6(\vec{r}_j)$ and the separation $r_{ij} = |\vec{r}_i - \vec{r}_j|$ for all domain pairs in the image giving $\frac{1}{2}N(N-1)$ data points, where N is the total number of bubbles in the image. Then we sort the data by the separation r_{ij} and average both, r_{ij} and $g_6(r_{ij})$ over blocks of size $N/2$. In figure 5.15b we plot $|\langle g_6(r_{ij})\rangle_{46}|$ against $\langle r_{ij}\rangle_{46}$ since we have 92 domains in image 5.12b. We observe that already at $r = 0$ we have a reduced correlation meaning that the local neighbourhood of each domain deviates significantly from a hexagon. Then, up to $r \approx 4$ μm there are no neighbours. The peak around 5 μm corresponds to the nearest neighbours for which some orientational correlation is still present. For larger separations the correlation rapidly decreases to zero.

It seems therefore clear that the bubble domains are not arranged on a regular triangular lattice. This is in contrast to the results found in simulations for the packing of hard disks [35] or particles with short-range Lennard-Jones interactions [36]. In both cases an initially disordered array of identical particles was found to quickly order when being annealed. Disorder was however found to occur in mixtures of hard spheres with different diameters or for Lennard-Jones particles with different interaction strength. Since the bubble domains in our case are not identical, in fact they are not even perfectly circular, some disorder may be expected. Another difference results from the fact that the bubbles interact via the long-ranged $1/r^3$ dipolar interaction. In the theoretical work by Muratov [29] a long-range interaction of Coulomb type $(1/r)$ was found to inevitably produce disorder in the domain pattern.

6 Non-equilibrium properties

In the previous chapters we have analysed equilibrium properties of the Fe/Cu(001)-films at variable temperatures and magnetic fields. We have restricted ourselves to regions of T-H-space where the discussed observables are *at equilibrium*, i.e. we truly expect them to represent the lowest (free) energy state. In this chapter we want to look into the dynamic response of the system to an external stimulus such as a change in temperature or applied field and the metastable states these kinetics may lead to.

6.1 Metastable domain patterns

6.1.1 Disordered domain pattern at low temperature

As anticipated in section 4.2, the first images of a measurement series appear random in the sense that they are not reproducible. After evaporation the sample is transferred from the evaporation stage to the measurement chamber and is fixed in the thermostat sample holder. During the transfer we have no control over the temperature and the field the sample is exposed to. A common feature of the first images is that the domain size increases as a function of time, see Fig. 6.1. This can be explained in two ways:

i) The film's atomic structure relaxes after evaporation. Assuming that better crystallinity leads to a higher T_C, this implies a decrease of the effective temperature T/T_C.

ii) The sample temperature after evaporation is high and the sample starts to cool down when the good thermal contact with the sample holder of the cryostat is established, also leading to a decrease of the effective temperature T/T_C.

In both cases, the initial stage of sample evolution can be viewed as the response to a decrease of the effective temperature. The evolution of the domain pattern at constant temperature (T=298 K) is shown in Fig. 6.1. The measurement of image **a** is started 30 minutes after the end of the film growth and 20 minutes after thermal contact with the sample holder has been established. From calibration measurements, see section 3.2.2, we expect therefore that the sample temperature really has reached room temperature. However, in image **a** the domain pattern shows considerable stripe-order with an average stripe width of $L_0 = 3.15 \pm 0.13$ μm corresponding to the equilibrium value at T=329 K (T/T_C=0.98). The relatively small error in L_0, determined from the width of the peak in the structure factor (inset) is an indication of the order. Notice that even the second-order peaks are faintly visible in the Fast-Fourier Transform (FFT) image. Some bubble-like, round domains are also found in the pattern. When

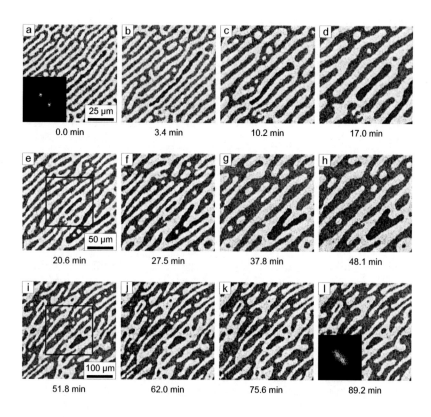

Figure 6.1: Initial coarsening of the domain pattern at constant sample holder temperature of 298 K (d=2.28 ML, T/T_C=0.89). The measurement of image **a** starts 30 minutes after the end of the film growth and 20 minutes after thermal contact with the sample holder has been established. The red squares in images **e** and **i** mark the field of view of images **d** and **h** respectively. Two black domains have been coloured blue to aid following the pattern transformation. The insets in images **a** and **l** show the modulus of their 2D Fourier transform. The residual field is compensated, i. e. the perpendicular field the sample is exposed to is smaller than 2 μT.

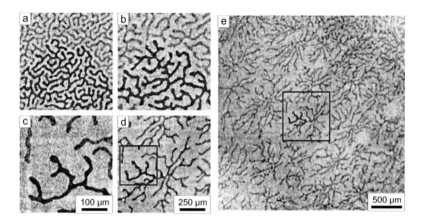

Figure 6.2: Coarsening of the domain pattern upon relaxing in the residual field, $\mu_0 H = -11.5$ µT, at a constant temperature of 305 K. The same domain is coloured blue in all images. Images **b** and **c** are measured 32 and 192 minutes after image **a** respectively. Image **d** shows a larger area and image **e** shows an overview (3 mm × 3 mm) of most of the total sample area (4 mm × 4 mm). The blurring of image e at the right and left borders is due to the limited focus depth of the microscope.

comparing images a to d, the domain width increases while the average domain length decreases, see e. g. the domain marked in blue. The small, circular domains collapse, therefore the total number of domains is decreasing monotonously. By these processes, the domain pattern acquires an increasingly disordered structure, apparently confirming the picture that the system is quenched to some random state. In image l the average domain size is $L_0 = 30 \pm 16$ µm. The large error is due to the broad peak in the structure factor (inset) reflecting the low degree of order present in the pattern. However, some memory of the original stripe orientation is retained, as is evident from the anisotropy of the FFT image. The fact that the low-temperature pattern observed after the initial coarsening is different for every sample rules out the possibility that the final stationary pattern (here image l) represents some weird equilibrium state.

6.1.2 Field cooled domain pattern

The situation is different if the sample is exposed to a small magnetic field while cooling. In this case, the external field compresses one type of domains and the coarsening of the pattern is asymmetric, see figure 6.2. Image a shows a labyrinth pattern that seems approximately symmetric with respect to black and white domains. One black domain extending over most of the field of view is coloured blue. Half an hour later (image b) the black domains have shrunken. Three hours later most of the domain branches have disappeared and the remaining black domain is star-shaped (images c, d) on a homogeneous white background. From image e it is evident that this is true for the entire sample. Such star-shaped domains have been observed as the result of domain growth upon demagnetizing a saturated sample [12,63,64]. Here we apparently follow the reverse process.

6.1.3 Labyrinthine pattern

The response of the sample to an increase in temperature is investigated in figure 6.3. After acquisition of image 6.1l the sample is heated in a step to 320 K ($T/T_C = 0.955$) and a waiting time of 4 minutes is allowed for the sample temperature, heating power and thermal drift to settle. Then the measurement of 6.3a is started. We extract a domain size $L_0 = 19 \pm 2$ µm, the smaller error being an indication that some order has been restored. This becomes more evident from comparing the structure factors, insets in Fig. 6.1l and Fig. 6.3a.

One particular white domain has been coloured yellow to allow tracking its evolution. After 30 minutes at 320 K (Fig. 6.3b), the domain pattern has reached a steady state with $L_0 = 12.1 \pm 0.6$ µm. In [7] it was shown that the domain size decreases with increasing temperature. For the domain size to decrease it is necessary to increase the domain wall density, i.e. to introduce additional domains into the pattern. One possibility is that a new domain nucleates somewhere in the sample and that it then elongates to form a stripe. Especially, the domain may nucleate at the sample border and then grow across the sample. We have however seen in section 5.1 that nucleation is not observed but at high temperatures. The dominant process observed is a transverse instability [6,15,29,65,66] leading to new domain branches growing laterally from the existing domains. The resulting pattern is more isotropic as can be seen in the FFT of image f that shows a ring of well-defined radius but considerable spectral weight for all angles. The original stripe orientation is however still dominant. At higher temperatures the stripe order is recovered, see image i, j, k.

The labyrinthine pattern is always observed when the sample is heated. This reproducibility rises the question whether it is the equilibrium pattern in an intermediate range of temperatures. Although this can not be excluded, theoretical considerations [29,67] show that a transverse instability is only expected if the actual domain size is *larger* than the equilibrium one. This is obviously the case upon heating but not upon cooling. An analysis of the mean-field Hamiltonian with respect to rotational symmetry [6] also showed that in zero field a pattern with two-fold symmetry, i. e. perfect stripes, has the lowest free energy at all temperatures. In agreement with these predictions but in contrast to previous work [6,15] we have never observed a transverse instability upon cooling the sample. This does however not imply that stripe order is necessarily preserved upon cooling, as is evident from the disorder induced by the coarsening of the domain pattern in Fig. 6.1.

As a possible explanation for the discrepancy we notice that a decreasing absolute

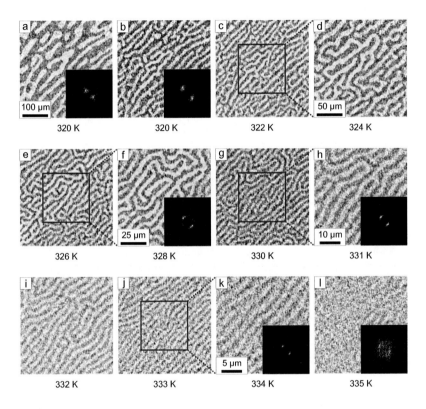

Figure 6.3: Pattern transformation upon heating in zero field (d=2.28 ML). The measurement of image **a** starts after a step-increase of the sample temperature from room temperature (Fig. 6.1l) to 320 K and a waiting time of 4 minutes. One particular domain is coloured yellow to follow the domain evolution. The insets show the modulus of the FFT of the corresponding images, the red squares mark the field of view of the subsequent image. A scale bar is indicated whenever the magnification is changed.

temperature is not always equivalent to a decreasing reduced temperature. Towards the end of the sample life time, the Curie temperature is often observed to decrease due to surface contaminants. If the evolution of the domain pattern is not followed continuously and after a long time a labyrinth is observed, it may be due to this decrease of T_C, leading at constant or even slightly decreasing absolute temperature to an effective increase of the reduced temperature, promoting a decrease of the domain size and the associated transverse instability.

From the different pattern evolutions after cooling (Fig. 6.1) and upon heating (Fig. 6.3) it is evident that at least one of the two, probably both, are not equilibrium processes. In consequence, possibly not all observables of the system are well described by equilibrium thermodynamics.

6.1.4 Magnetic hysteresis

As is obvious from Fig. 6.4, at low temperatures the Fe-films exposed to a changing applied field show considerable hysteresis, the magnetization $\overline{m}$ may differ significantly from the equilibrium value corresponding to a certain field. For not too high values of the applied field, the change in magnetization proceeds only by reversible domain wall propagation. As a result of such processes, at high enough applied fields the sample is mostly magnetized homogeneously along the applied field, except for a network of narrow inverted domains [68–71] whose stability against collapsing is explained by the high field required to eliminate a domain of striped shape, see section 2.6. When having a closer look at the transition at constant applied field from image c to d in the time domain, (Fig. 6.4j) we observe that $\overline{m}(t)$ follows an exponential relaxation,

$$\overline{m}(t) = \overline{m}_\infty + (\overline{m}_0 - \overline{m}_\infty) \exp\left(-\frac{t - t_0}{\tau}\right) , \qquad (6.1)$$

characterized by a time scale τ. Such a relaxation is expected when the excitation frequency is considerably below the resonance of the domain walls in their pinning potentials [72]. In this case, the change of the magnetization is proportional to its deviation from equilibrium,

$$\frac{\partial \overline{m}}{\partial t} = -\frac{1}{\tau}\left(\overline{m}(t) - \overline{m}_\infty\right) \qquad (6.2)$$

The same behaviour is observed between images g and h, as shown in Fig. 6.4k. The observed characteristic times are 14.5 minutes (j) and 11.5 minutes (k).

Figure 6.5a shows the details of the domain wall motion in the pattern transformation from image 6.4b with only narrow white domains to 6.4d with only narrow black domains. From the time series the important role of pining sites is evident: At the sites marked with red circles the domain wall stays attached during the entire process. Between these pinning sites the domains grow laterally [55].

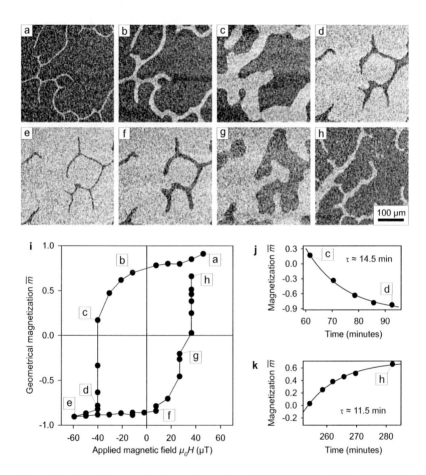

Figure 6.4: Magnetic hysteresis at room temperature (T=298 K, T/T_C=0.90, d=2.15 ML). The images **a** to **h** show selected stages in a magnetization reversal process summarized in the hysteresis loop **i**. The plots **j** (**k**) show the slow magnetic relaxation at constant field, observed between images **c** and **d** (**g** and **h**), in the time domain. The indicated times refer to the elapsed minutes since the start of the measurement (image **a**).

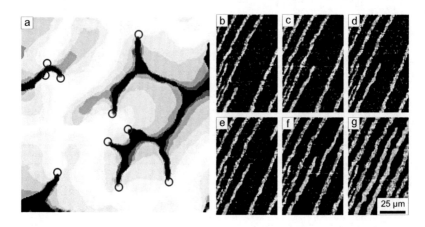

Figure 6.5: a Superimposed black-white images illustrating the lateral domain growth process leading from image 6.4b to 6.4d. Red circles mark pinning sites at which the domain wall remains attached throughout the series. **b** to **g** Longitudinal creep-like domain growth after a field pulse (+700 µT). (d=3.2 ML, T=297 K) The applied fields are: -12 µT for images **b** to **e**, -60 µT for **f** and -110 µT for **g**. Times are t_b=0 min, t_c=7 min, t_d=15 min, t_e=29 min, t_f=36 min, t_g=43 min.

A different growth mode is observed in Fig. 6.5b to g. A field pulse of +700µT is applied to the sample producing an almost saturated ($\overline{m} \approx 1$) state. After removal of the applied field, image 6.5b is observed in the uncompensated residual field ($\mu_0 H$ = -12 µT). Some stripe domains are still present. As time elapses, white stripe domains growing from outside the field of view, coloured in yellow, fill the black space created by the field pulse. The strikingly different magnetization process may be due to larger film thickness (3.2 ML) or a different distribution of pinning centres on the sample. It may also be a result of the domain pattern itself. The residual domains in Fig. 6.4 are disordered and approximately isotropic while in Fig. 6.5b the remaining stripe domains clearly establish a preferred domain orientation. Notice also that the stripe domains in the latter case are only a few microns wide, while the homogeneous domains in Fig. 6.4 extend over much larger distances.

6.1.5 Field response vs. temperature

Figure 6.6 shows four images taken at $T = 300$ K (T/T_C=0.94) at the magnetic fields and times as indicated. We have $\overline{m}_a = 0$, $\overline{m}_b = 0.36$, $\overline{m}_c = -0.24$ and $\overline{m}_d = 0$. Although the amplitude of the magnetic field change is similar to (Fig. 6.4), no hysteresis can be detected, the measurement appears quasi-static. This is an indication

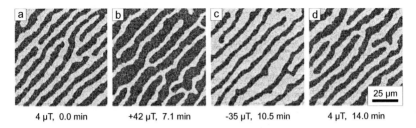

| 4 µT, 0.0 min | +42 µT, 7.1 min | -35 µT, 10.5 min | 4 µT, 14.0 min |

Figure 6.6: Response to weak applied fields at intermediate temperature. (d=1.89 ML, T=300 K, T/T_C=0.94)

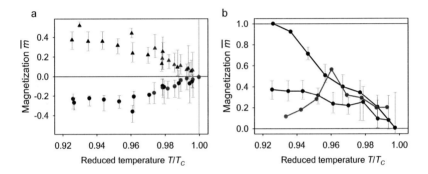

Figure 6.7: a Induced magnetization $\overline{m}$ in response to a weak applied DC-field of +42 µT (triangles) and -35 µT (circles). Different colors correspond to different measurement series. **b** Induced magnetization $\overline{m}$ upon cooling in a constant applied field of +42 µT (blue), in response to a DC-applied field of +42 µT at constant temperature (black) and attained in AC-measurements with 42 µT amplitude and a maximum period of 20 seconds.

that the relaxation is considerably faster than the image acquisition. However, the magnitude $\overline{m}_{\mathrm{max}}$ of the response to the field is smaller than for the hysteretic low-T case. The sample consists of relatively well ordered stripe domains that respond to the applied field mainly by a change of the stripe width, essentially without changing the position of the domains. In order to quantify the temperature dependence of $\overline{m}_{\mathrm{max}}$ we investigate the DC-response of the system to a step-change in the applied field of constant amplitude in figure 6.7. Figure **a** shows the magnetization $\overline{m}$ induced by a weak applied field of +42 µT (triangles) and -35 µT (circles). The different colours correspond to different measurement series. At high temperatures the absolute value of $\overline{m}$ increases with decreasing temperature. Since the applied field is low, the susceptibility is linear and we can write $\chi = \frac{\partial \overline{m}}{\partial H} \approx \frac{\overline{m}}{H}$. In the theory part 2.5 we have seen that the susceptibility scales with L_0/M_S, and the increase of χ with decreasing T/T_C can be attributed to the increasing L_0. From comparison of the different measurement series we see that this behaviour is quite reproducible. Below about $0.96\,T_C$ however, a saturation of the induced $\overline{m}$ is observed. In figure b the results of Fig. a (black) are compared to $\overline{m}(T)$ obtained from cooling in a constant field of $+42\,\mu T$ (blue) and the asymptotic value $\overline{m}_\infty$ attained in an AC-measurement with $+42\,\mu T$-amplitude and a maximum period of 20 seconds (red). Above about $0.96\,T_C$ the values obtained by the different methods agree within their error. At lower temperatures however, time seems to play a crucial role. The response is largest for the sample cooled in the field, it increases monotonously for decreasing temperature. The response to the field step reaches a constant value $\overline{m}$ between 0.3 and 0.4. The response in the relatively fast AC-mode decreases as the temperature is decreased.

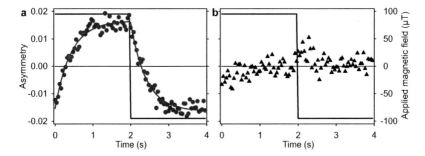

Figure 6.8: SEMPA-AC-measurement of the magnetization. The magnetic field is applied as a square wave with period 4 seconds and amplitude 95 µT. Figure **a** shows the response of the average out-of-plane Mott-asymmetry (red dots) to the applied field (black solid line). The response to each slope of the field is well fitted by an exponential, eq. (6.1), red solid line. Figure **b** shows the in-plane Mott-asymmetry, no response to the field is observed.

6.2 Relaxation times

At low temperatures, where the relaxation times are larger or comparable to the acquisition time of an image (minutes), τ can be determined from image sequences, as in Fig. 6.4. At higher temperatures however, the relaxation times are expected to decrease and reach the time scales of one scan-line (seconds), one image pixel (milliseconds) or even less. In this regime a different measurement scheme needs to be adopted.

6.2.1 AC-measurements with SEMPA

As described in section 3.2.4, the SEMPA can be run in a special AC-mode. The result of such a measurement is shown in Fig. 6.8. The magnetic field (black solid curve) is applied as a square wave of period 4 s and amplitude 95 µT. For each period 100 samples are taken and a total of 50 loops is recorded, giving an integration time of 2 s per data point. Figure **a** shows the out-of-plane component of the magnetization with two independent exponential fits (6.1), giving characteristic times $\tau_{\mathrm{rise}} = 425 \pm 35$ ms and $\tau_{\mathrm{fall}} = 408 \pm 33$ ms. In Fig. b, showing the in-plane component of the magnetization, no response to the applied field is observed. The exponential fit to the relaxation data not only gives the relaxation time τ but also the constant magnetization attained after long enough time, $\overline{m}_{\infty}$. The average Mott asymmetry A measured in the AC-mode is the product of the geometric magnetization $\overline{m}$ and the saturation magnetization M_S. However, from images acquired in DC applied field also at high temperatures we see that the weak field does not affect M_S. Therefore we conclude that the change observed in A is due to the domain wall motion leading to a change in $\overline{m}$. Together with

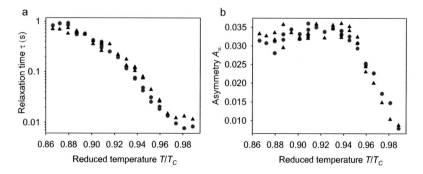

Figure 6.9: a Relaxation times τ_{rise} (blue) and τ_{fall} (red) vs. reduced temperature in a log-linear plot. (d=2.14 ML, T_C=344 K). The time constants are extracted from AC-measurements with a square-wave excitation with constant amplitude ±95 µT. The period of the excitation is adapted to the relaxation times in order to cover a range of ca. 10τ. **b** Stationary Mott-asymmetry A_∞ as extracted from the same AC-measurements in response to rising (blue) and falling field (red).

the universal curve for $M_S(T)$ (4.1) we can therefore extract indirectly the stationary displacement of $\overline{m}$, $\overline{m}_\infty = A_\infty/M_S(T)$ also from the AC-data.

6.2.2 Relaxation time vs. temperature

We now want to get a quantitative insight into the temperature dependence of the response to a step-like change in the applied field. For this purpose, we repeat measurements similar to Fig. 6.8 at different temperature. Figure 6.9a shows the extracted relaxation times from AC-measurements with a constant excitation amplitude of ±95 µT and period adapted to the relaxation time. At the highest temperatures ($T/T_C > 0.97$) the relaxation times lie below the reliably accessible time resolution of ≈ 10 ms and at low temperatures ($T/T_C < 0.90$) the relaxation times seem to saturate at around 1 second. In an intermediate temperature range the relaxation times show an approximate linear behaviour in the log-linear plot of Fig. 6.9a, indicating an exponential increase with decreasing temperature. Figure 6.9b shows the temperature dependence of the stationary magnetization $\overline{M}_\infty$ attained before the reversing field is applied. It shows an approximately linear decrease for $T/T_C > 0.94$ extrapolating to zero at T_C. In a range of intermediate temperatures $\overline{M}_\infty$ becomes constant at approximately M_S. In this range, the sample essentially switches between $\overline{m}_\infty \approx \pm 1$. At low temperatures the excited amplitude decreases as is also observed in measurements with lower excitation amplitude, see figure 6.7b.

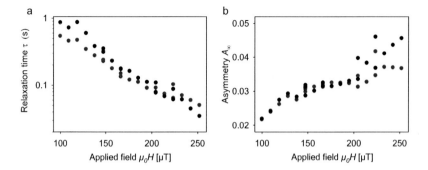

Figure 6.10: a Relaxation times τ_{rise} (blue) and τ_{fall} (red) vs. field amplitude at constant temperature. (T=300 K, d=2.14 ML). **b** Stationary Mott-asymmetry A_∞ as extracted from the same AC-measurements in response to rising (blue) and falling field (red).

6.2.3 Relaxation time vs. field amplitude

The dependence of the relaxation time on the applied field is investigated in figure 6.10a. The temperature is kept constant at T=300 K. A strong increase of τ with decreasing field is observed. The asymmetry (Fig. 6.10a) is roughly constant at the saturation value for fields between 150 and 200 µT. Beyond 200 µT the measurement less reliable because the shift in the field of view caused by the deflection of the SEM-beam in the applied field becomes important and the secondary electrons are deflected strongly on their way to the detector. At lower fields, the excited amplitude $\overline{m}_\infty$ becomes smaller, indicating that the response of the system decreases, at least on the time scale observable in AC mode. Figure **a** suggests an exponential increase of τ with decreasing H, in general agreement with the model put forward by Venus and Dunlavy [72]

$$\tau \propto \exp\left(\frac{\pm\mu_0 H M_S V_B}{k_B T}\right) \tag{6.3}$$

where the sign depends on whether the applied field helps or hinders the relaxation and V_B is the Barkhausen volume that is reversed as the domain wall moves from one pinning site to the next. From a fit to the data in Fig. 6.10a we obtain $\frac{M_S V_B}{k_B T} \approx$ 0.02 µT^{-1}. Using T=300 K and M_S=2·10^6 A/m we obtain then $V_B \approx 4\cdot10^{-24}$ m^3 corresponding to a sample area of roughly 90 nm by 90 nm. A value that seems reasonable.

6 *Non-equilibrium properties*

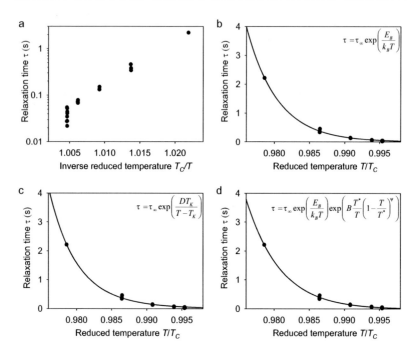

Figure 6.11: **a** Arrhenius plot $\log(\tau)$ vs. $1/T$ of the relaxation times measured in zero field. **b** Fit to the data of **a** using an Arrhenius law as indicated in the figure. **c** Fit to the data using a Vogel-Fulcher law as indicated. **d** Fit to the data using modified Arrhenius law with $E_B(T)$ as proposed by Kivelson and coworkers, the function is indicated in the figure.

6.2.4 Relaxation in zero applied field

In the previous subsections we have investigated the relaxation times in response to an applied field, i. e. the field-driven domain wall motion. At zero applied field any state with $\overline{m} \neq 0$ is an excited state and should in principle relax to a state with $\overline{m} = 0$. Since the corresponding time scales become large at low temperature, this relaxation is only observable at high temperatures, close to T_C. Figure 6.11a shows the relaxation times in an Arrhenius plot, $\ln(\tau)$ vs. $1/T$. In this plot, the functional dependence is approximately linear, indicating that the relaxation is indeed thermally activated. A fit to the data using an Arrhenius law (Fig. 6.11b) gives $\tau_\infty \approx 2 \cdot 10^{-93}$ s and $E_B \approx 216 k_B T_C$. The value for the relaxation time at high temperatures τ_∞ obtained from this fit is very unphysical.

The temperature dependence of the relaxation time in a frustrated, glassy, system is non-Arrhenius like [34] and various forms have been proposed to account for this fact, as has been discussed in section 2.7.2. Fig. 6.11c shows a fit to the data using a Vogel-Fulcher formula, giving the parameters $\tau_\infty \approx 6 \cdot 10^{-11}$ s, $D \approx 5$ and $T_K/T_C \approx 0.85$. Notice that the quality of the fit and the fit itself are indistinguishable from the Arrhenius law, but the value we obtain for τ_∞ in the picosecond region is physically reasonable. Kivelson and collaborators [73] have suggested a modified Arrhenius law with a temperature dependent barrier to account for the large increase of the relaxation time. A fit to their law as shown in Fig. 6.11d, yields the parameters $\tau_\infty \approx 6 \cdot 10^{-10}$ s, $E_B \approx 23k_BT_C$, $B \approx 90$ and $\psi \approx 0.75$. Again, the fit is indistinguishable from the other two but compared to the bare Arrhenius law τ_∞ seems more reasonable.

We can therefore state that the relaxation time behaves as

$$\tau = \tau_\infty \exp\left(\frac{E_B(T)}{k_BT} \pm \frac{\mu_0 H M_S V_B}{k_BT}\right) . \tag{6.4}$$

Although we find the increase of $E_B(T)$ with decreasing T only indirectly, we may take this result as a preliminary indication that super-Arrhenius-like behaviour is indeed present in the ultrathin Fe-films investigated here, providing a starting point for future experiments.

Appendix

A Energy calculations

A.1 Magnetostatic energy

Considering a sample as described in section 2.2, the demagnetizing energy or magnetostatic self-energy [2] can be written as

$$E_{\mathrm{MS}} = -\frac{\mu_0}{2} \int_V \vec{M}(\vec{r}) \cdot \vec{H}_{\mathrm{Demag}}(\vec{r}) \, \mathrm{d}^3 V \tag{A.1}$$

Where H_{Demag} is the magnetic field produced by the magnetic sample itself. The demagnetizing field can be calculated from the magnetic charge $\rho_m = -\vec{\nabla} \cdot \vec{M}$ via a magnetic scalar potential $\Phi_m(\vec{r})$ [1,2,6].

$$\vec{H}_{\mathrm{Demag}} = -\vec{\nabla}\Phi_m(\vec{r}) \quad \text{with} \quad \triangle\Phi_m(\vec{r}) = -\rho_m \tag{A.2}$$

The scalar potential is then given by

$$\Phi_m(\vec{r}) = \frac{1}{4\pi} \int_{V'} \frac{\rho_m(\vec{r}')}{|\vec{r} - \vec{r}|} \, \mathrm{d}^3 V \tag{A.3}$$

and the magnetic charge ρ_m is

$$\rho_m = -\vec{\nabla} \cdot \vec{M}(\vec{r}) \qquad \text{inside the sample and} \tag{A.4}$$

$$\rho_m = \vec{n} \cdot \vec{M}(\vec{r}) \qquad \text{at the surface of the sample} \tag{A.5}$$

where $\vec{n}$ is the unit normal vector pointing out of the sample. In our case $\vec{M} \parallel \vec{e}_z$ and $\frac{\partial M}{\partial z} = 0$ inside the sample so that the magnetic charge is restricted to the top and bottom surfaces of the film.

$$\rho_m(\vec{r}') = M(x', y') \left[\delta(z' - d) - \delta(z') \right] \tag{A.6}$$

With $\vec{H}_{\mathrm{Demag}} = -\vec{\nabla}\Phi$ eq. (A.1) becomes

$$E_{\mathrm{MS}} = \frac{\mu_0}{2} \int_V \vec{M}(\vec{r}) \cdot \vec{\nabla}\Phi(\vec{r}) \, \mathrm{d}^3 V \tag{A.7}$$

Integrating by parts gives

$$E_{\mathrm{MS}} = \frac{\mu_0}{2} \int_{\partial V} \vec{M}(\vec{r}) \cdot \vec{n} \, \Phi(\vec{r}) \, \mathrm{d}^2 S - \frac{\mu_0}{2} \int_V \vec{\nabla} \cdot \vec{M}(\vec{r}) \Phi(\vec{r}) \, \mathrm{d}^3 V \tag{A.8}$$

$$= \frac{\mu_0}{2} \int_V \rho_m(\vec{r}) \Phi(\vec{r}) \, \mathrm{d}^3 V \tag{A.9}$$

The surface term vanishes because the surface ∂V of the integration volume lies outside the sample where $\vec{M} = 0$. The magnetostatic energy reduces thus to the interaction between the magnetic (surface) charges:

$$E_{\mathrm{MS}} = \frac{\mu_0}{2} \int_{S1} \rho_m(\vec{r}_1) \Phi_1(\vec{r}_1)\, \mathrm{d}^2 S_1 + \frac{\mu_0}{2} \int_{S1} \rho_m(\vec{r}_1) \Phi_2(\vec{r}_1)\, \mathrm{d}^2 S_1$$
$$+ \frac{\mu_0}{2} \int_{S2} \rho_m(\vec{r}_2) \Phi_1(\vec{r}_2)\, \mathrm{d}^2 S_2 + \frac{\mu_0}{2} \int_{S2} \rho_m(\vec{r}_2) \Phi_2(\vec{r}_2)\, \mathrm{d}^2 S_2 \tag{A.10}$$

$$= \mu_0 \int_{S1} \rho_m(\vec{r}_1) \Phi_1(\vec{r}_1)\, \mathrm{d}^2 S_1 + \mu_0 \int_{S1} \rho_m(\vec{r}_1) \Phi_2(\vec{r}_1)\, \mathrm{d}^2 S_1 \tag{A.11}$$

where indices 1 and 2 refer to the top and bottom surfaces of the film respectively. The last line reads explicitly

$$E_{\mathrm{MS}} = \mu_0 \int_{S1} \rho_m(\vec{r}) \frac{1}{4\pi} \int_{S1} \frac{\rho_m(\vec{r}')}{|\vec{r}' - \vec{r}|}\, \mathrm{d}^2 S'\, \mathrm{d}^2 S + \mu_0 \int_{S1} \rho_m(\vec{r}) \frac{1}{4\pi} \int_{S2} \frac{\rho_m(\vec{r}')}{|\vec{r}' - \vec{r}|}\, \mathrm{d}^2 S'\, \mathrm{d}^2 S \, . \tag{A.12}$$

We introduce the 2-dimensional vector $\vec{\rho} = (x, y)$ for the in-plane component of $\vec{r}$ and notice that $\rho_m(\vec{r})$ on the top surface ($S1$) is simply given by $M(\vec{\rho})$ while on the bottom surface $S2$ the magnetic charge is given by $-M(\vec{\rho})$. Using this we can write

$$E_{\mathrm{MS}} = \mu_0 \int_S M(\vec{\rho}) \frac{1}{4\pi} \int_S \frac{M(\vec{\rho}')}{|\vec{\rho}' - \vec{\rho}|}\, \mathrm{d}^2 S'\, \mathrm{d}^2 S - \mu_0 \int_S M(\vec{\rho}) \frac{1}{4\pi} \int_S \frac{M(\vec{\rho}')}{\sqrt{|\vec{\rho}' - \vec{\rho}|^2 + d^2}}\, \mathrm{d}^2 S'\, \mathrm{d}^2 S \tag{A.13}$$

$$= \frac{\mu_0}{4\pi} \int_S M(\vec{\rho}) \int_S M(\vec{\rho}') \left(\frac{1}{|\vec{\rho}' - \vec{\rho}|} - \frac{1}{\sqrt{|\vec{\rho}' - \vec{\rho}|^2 + d^2}} \right)\, \mathrm{d}^2 S'\, \mathrm{d}^2 S \tag{A.14}$$

Where d is the thickness of the magnetic layer. Let's have a closer look at the interaction potential

$$V_d(|\vec{\rho}' - \vec{\rho}|) = \frac{1}{|\vec{\rho}' - \vec{\rho}|} - \frac{1}{\sqrt{|\vec{\rho}' - \vec{\rho}|^2 + d^2}} \, . \tag{A.15}$$

In this work we are interested in the case of small d. We substitute $x = |\vec{\rho}' - \vec{\rho}|$ and obtain

$$V_d(x) = \frac{1}{x} - \frac{1}{x\sqrt{1 + d^2/x^2}} \, . \tag{A.16}$$

For small $\varepsilon = \frac{d}{x}$ we may expand $V_d(x)$ as a Taylor series around $\varepsilon = 0$. We obtain

$$V(x, \varepsilon)\big|_{\varepsilon=0} \qquad\qquad\qquad\qquad\qquad\qquad\qquad\qquad = 0 \tag{A.17}$$

$$\frac{\partial V(x, \varepsilon)}{\partial \varepsilon}\bigg|_{\varepsilon=0} = \frac{\varepsilon}{x\left(1 + \varepsilon^2\right)^{\frac{3}{2}}}\bigg|_{\varepsilon=0} \qquad\qquad\qquad\qquad = 0 \tag{A.18}$$

$$\frac{\partial^2 V(x, \varepsilon)}{\partial \varepsilon^2}\bigg|_{\varepsilon=0} = \frac{1}{x\left(1 + \varepsilon^2\right)^{\frac{3}{2}}}\bigg|_{\varepsilon=0} - \frac{3\varepsilon^2}{x\left(1 + \varepsilon^2\right)^{\frac{5}{2}}}\bigg|_{\varepsilon=0} \qquad = -\frac{1}{x} \, . \tag{A.19}$$

To leading order in d/x the interaction potential therefore becomes

$$V_d(x) \approx \frac{1}{2}\frac{1}{x}\frac{d^2}{x^2} = \frac{1}{2}\frac{d^2}{x^3}\,. \tag{A.20}$$

leading to the magnetostatic energy in dipole approximation

$$E_{\mathrm{Dip}} = \frac{\mu_0}{8\pi}\int_S\int_S \frac{M(\vec{\rho})M(\vec{\rho'})d^2}{|\vec{\rho'}-\vec{\rho}|^3}\,\mathrm{d}^2S'\,\mathrm{d}^2S\,. \tag{A.21}$$

Note that the above expression has a divergence for small $|\vec{\rho'}-\vec{\rho}|$. However, in that limit we no longer have $\varepsilon \ll 1$ and must therefore go back to (A.15) where the order of the divergence is r^{-1} rather than r^{-3}, making it integrable in 2 dimensions. The apparent divergence of the r^{-1}-potential (A.15) at large r is lifted because for large r $V_d(r)$ falls off like r^{-3} as we have just seen. The total magnetostatic energy is therefore well-defined and finite.

As an example we compute the total magnetostatic energy per unit volume of a laterally infinite, homogeneously magnetized slab.

$$e_{\mathrm{MS,hom}} = \frac{1}{d(2\Lambda)^2}\frac{\mu_0}{4\pi}M^2\int_S\int_S \frac{1}{|\vec{\rho'}-\vec{\rho}|} - \frac{1}{\sqrt{|\vec{\rho'}-\vec{\rho}|^2+d^2}}\,\mathrm{d}^2S'\,\mathrm{d}^2S \tag{A.22}$$

$$= \frac{1}{d}\lambda\int_S \frac{1}{x} - \frac{1}{\sqrt{x^2+d^2}}\,\mathrm{d}^2S' \tag{A.23}$$

where we have introduced x as before, λ as defined in (2.20) and have used that due to translation invariance the outer integral simply gives the sample area $(2\Lambda)^2$. We go to polar coordinates

$$e_{\mathrm{MS,hom}} = \frac{\lambda}{d}\lim_{\Lambda\to\infty}\int_0^\Lambda\int_0^{2\pi}\left(\frac{1}{x}-\frac{1}{\sqrt{x^2+d^2}}\right)x\,\mathrm{d}\varphi\,\mathrm{d}x \tag{A.24}$$

$$= \frac{2\pi\lambda}{d}\lim_{\Lambda\to\infty}\int_0^\Lambda\left(1-\frac{x}{\sqrt{x^2+d^2}}\right)\mathrm{d}x \tag{A.25}$$

$$= \frac{2\pi\lambda}{d}\lim_{\Lambda\to\infty}\left[x-\sqrt{x^2+d^2}\right]_{x=0}^\Lambda \tag{A.26}$$

$$= \frac{2\pi\lambda}{d}\left(\lim_{\Lambda\to\infty}\left(\Lambda-\sqrt{\Lambda^2+d^2}\right)+d\right) \tag{A.27}$$

$$e_{\mathrm{MS,hom}} = 2\pi\lambda \tag{A.28}$$

Starting from (A.14) we want to find the Fourier representation of E_{MS}. We use the expression [6, 20]

$$\frac{1}{|\vec{r}_i-\vec{r}_j|} = \frac{1}{2\pi}\iint \frac{e^{i\vec{q}_\parallel\cdot(\vec{\rho}_i-\vec{\rho}_j)}\cdot e^{-q_\parallel|z_i-z_j|}}{q_\parallel}\,\mathrm{d}^2q_\parallel \tag{A.29}$$

where $\vec{q}_{\|} = (q_x, q_y)$ is the in-plane $\vec{q}$-vector and $q_{\|}$ its modulus. For simplicity we set $q := q_{\|}$. With this, the scalar potential becomes

$$\Phi_m(\vec{r}) = \frac{1}{8\pi^2} \int_{V'} \rho_m(\vec{r}') \iint \frac{e^{i\vec{q}\cdot(\vec{\rho}_i - \vec{\rho}_j)} \cdot e^{-q|z_i - z_j|}}{q} \, \mathrm{d}^2 q \, \mathrm{d}^3 V \tag{A.30}$$

and the magnetostatic energy

$$E_{\mathrm{MS}} = \frac{\mu_0}{8\pi^2} \int_S M(\vec{\rho}) \int_{S'} M(\vec{\rho}') \iint \frac{e^{i\vec{q}\cdot(\vec{\rho} - \vec{\rho}')}}{q} \, \mathrm{d}^2 q \, \mathrm{d}^2 S' \, \mathrm{d}^2 S \tag{A.31}$$

$$- \frac{\mu_0}{8\pi^2} \int_S M(\vec{\rho}) \int_{S'} M(\vec{\rho}') \iint \frac{e^{i\vec{q}\cdot(\vec{\rho} - \vec{\rho}')} \cdot e^{-qd}}{q} \, \mathrm{d}^2 q \, \mathrm{d}^2 S' \, \mathrm{d}^2 S \tag{A.32}$$

$$= \frac{\mu_0}{8\pi^2} \int_S M(\vec{\rho}) \int_{S'} M(\vec{\rho}') \iint \frac{e^{i\vec{q}\cdot(\vec{\rho} - \vec{\rho}')} \cdot \left(1 - e^{-qd}\right)}{q} \, \mathrm{d}^2 q \, \mathrm{d}^2 S' \, \mathrm{d}^2 S \tag{A.33}$$

We define the Fourier transform of $m(\vec{\rho})$ as

$$m(\vec{q}) = \frac{1}{2\pi} \iint m(\vec{\rho}) e^{-i\vec{q}\cdot\vec{\rho}} \, \mathrm{d}^2 \rho \tag{A.34}$$

and its inverse

$$m(\vec{\rho}) = \frac{1}{2\pi} \iint m(\vec{q}) e^{i\vec{q}\cdot\vec{\rho}} \, \mathrm{d}^2 q \tag{A.35}$$

Using these definitions, (2.6) and (2.20) the magnetostatic energy can be written as [6]

$$E_{\mathrm{MS}} = \frac{\mu_0(g\mu_B S)^2}{2a^6} \iint m(\vec{q}) \, m(-\vec{q}) \, \frac{1 - e^{-qd}}{q} \, \mathrm{d}^2 q \tag{A.36}$$

$$= 2\pi\lambda \iint m(\vec{q}) \, m(-\vec{q}) \, \frac{1 - e^{-qd}}{q} \, \mathrm{d}^2 q \tag{A.37}$$

A.2 Stripe domains - modulation along 1 dimension

We consider a striped system, i. e. $m(x,y)$ is independent of y and periodic along x with period $2L$. Without loss of generality we can choose the magnetization profile $m(x), x \in (-L, L)$ to be an even function of x and we can write

$$m(x,y) = m(x) = \sum_{n=0}^{\infty} a_n \cos(k_n x) \quad \text{with} \quad k_n = \frac{n\pi}{L} \tag{A.38}$$

and

$$a_0 = \frac{1}{2L} \int_{-L}^{L} m(x) \, \mathrm{d}x \tag{A.39}$$

$$a_n = \frac{2}{2L} \int_{-L}^{L} m(x) \cos\left(k_n \cdot x\right) \, \mathrm{d}x . \tag{A.40}$$

We rewrite (A.33) to suit this particular case

$$E_{\text{MS}} = \frac{\lambda}{2\pi} \iint m(x) \iint m(x') \iint \frac{e^{iq_x(x-x')}e^{iq_y(y-y')} \cdot \left(1 - e^{-qd}\right)}{q} \, \mathrm{d}q_x \, \mathrm{d}q_y \, \mathrm{d}x' \, \mathrm{d}y' \, \mathrm{d}x \, \mathrm{d}y$$

(A.41)

and use

$$\iint e^{iq_y(y-y')} \, \mathrm{d}y \, \mathrm{d}y' = 2\Lambda \, 2\pi \, \delta(q_y)$$

(A.42)

to arrive at

$$E_{\text{MS}} = 2\Lambda\lambda \int m(x) \int m(x') \iint \frac{e^{iq_x(x-x')} \cdot \left(1 - e^{-qd}\right)}{q} \delta(q_y) \, \mathrm{d}q_x \, \mathrm{d}q_y \, \mathrm{d}x' \, \mathrm{d}x$$

(A.43)

$$= 2\Lambda\lambda \int m(x) \int m(x') \int \frac{e^{iq_x(x-x')} \cdot \left(1 - e^{-|q_x|d}\right)}{|q_x|} \, \mathrm{d}q_x \, \mathrm{d}x' \, \mathrm{d}x$$

(A.44)

$$= 2\Lambda\lambda \int m(x) \int m(x') \int_0^\infty 2\cos(q_x(x-x')) \frac{1 - e^{-q_x d}}{q_x} \, \mathrm{d}q_x \, \mathrm{d}x' \, \mathrm{d}x$$

(A.45)

We now use the Fourier series for $m(x)$ (A.38) and obtain

$$E_{\text{MS}} = 2\Lambda d2\lambda \sum_{n=0}^\infty \sum_{m=0}^\infty a_n a_m \int \cos(k_n x) \int \cos(k_m x') \int_0^\infty \cos(q_x(x-x')) \frac{1 - e^{-q_x d}}{d\, q_x} \, \mathrm{d}q_x \, \mathrm{d}x' \, \mathrm{d}x$$

(A.46)

In the following we use the identities

$$\iint \cos(q_x(x-x')) \, \mathrm{d}x' \, \mathrm{d}x = \iint \cos(q_x x)\cos(q_x x') - \sin(q_x x)\sin(q_x x') \, \mathrm{d}x' \, \mathrm{d}x$$

$$= \iint \cos(q_x x)\cos(q_x x') \, \mathrm{d}x' \, \mathrm{d}x$$

(A.47)

and

$$\int \cos(k_n x)\cos(q_x x) \, \mathrm{d}x = \frac{1}{2} \int \cos((k_n - q_x)x) + \cos((k_n + q_x)x) \, \mathrm{d}x$$

$$= \pi \left(\delta(k_n + q_x) + \delta(k_n - q_x)\right)$$

(A.48)

because

$$\int_{-\Lambda}^{\Lambda} \cos(ax) \, \mathrm{d}x = 2\pi\delta(a) \,.$$

(A.49)

With

$$f(q_x) = \frac{1 - e^{-q_x d}}{d\, q_x}$$

(A.50)

we have after the x-integration

$$d \int \int_0^\infty \cos(k_m x') \cos(q_x x') \pi \big(\delta(k_n + q) + \delta(k_n - q)\big) f(q_x) \, dq_x \, dx' . \qquad (A.51)$$

For $q_x \geq 0$ and $k_n > 0$ the first delta function does not contribute to the q_x-integration. We will treat the $k_n = 0$ case separately later. For $k_n > 0$ we get

$$d\pi \int \cos(k_m x') \cos(k_n x') f(k_n) \, dx' . \qquad (A.52)$$

By performing the x'-integration we obtain

$$d\frac{\pi}{2} 2\Lambda \big(\delta_{k_m, -k_n} + \delta_{k_m, +k_n}\big) f(k_n) . \qquad (A.53)$$

The first Kronecker-δ gives no contribution because we set $k_n > 0$ before and $k_m \geq 0$ by definition. By carrying out the summation in (A.46) starting from $n = m = 1$ we obtain

$$E_{\mathrm{MS},1} = (2\Lambda)^2 d\pi\lambda \sum_{n=1}^\infty a_n^2 \frac{1 - e^{-k_n d}}{d \, k_n} \quad \text{with} \quad k_n = \frac{n\pi}{L} . \qquad (A.54)$$

We now have to consider the special case $k_n = 0$ in eq. (A.51)

$$E_{\mathrm{MS},2} = 2\Lambda d 2\lambda \sum_{m=0}^\infty a_0 a_m \int \cos(k_m x') \int_0^\infty 2\pi \delta(q_x) \cos(q_x x') \, f(q_x) \, dq_x \, dx' . \qquad (A.55)$$

We use that for an arbitrary even function $f(x)$

$$\int_0^\infty \delta(x) f(x) \, dx = \frac{1}{2} \int_{-\infty}^\infty \delta(x) f(x) \, dx = \frac{1}{2} f(0) \qquad (A.56)$$

and arrive at

$$E_{\mathrm{MS},2} = 2\Lambda d 2\lambda 2\pi \sum_{m=0}^\infty a_0 a_m \int \cos(k_m x') \frac{1}{2} f(0) \, dx' \qquad (A.57)$$

$$= 2\Lambda d 2\lambda 2\pi \sum_{m=0}^\infty a_0 a_m 2\Lambda \delta_{k_m, 0} \frac{1}{2} f(0) \qquad (A.58)$$

$$= (2\Lambda)^2 d\pi\lambda \, 2a_0^2 \qquad (A.59)$$

where we have used in the last step that $f(0) = 1$. By collecting the results (A.54) and (A.57) we arrive at the following expression for the dipolar energy of a striped system with period $2L$ along x:

$$E_{\mathrm{MS}} = (2\Lambda)^2 d\pi\lambda \left(2a_0^2 + \sum_{n=1}^\infty a_n^2 \frac{1 - e^{-k_n d}}{d \, k_n} \right) \quad \text{with} \quad k_n = \frac{n\pi}{L} \qquad (A.60)$$

Using this expression we want to compute the ground state energy of a striped system in a magnetic field. At zero temperature in the strict Ising case the domain boundary is atomically sharp with $\sigma_w = 2J$, where J is the exchange coupling constant. The magnetization profile along x is therefore a square wave and we define $m(x,y) = m(x)$ as an even function of x, and call the smaller stripe width w. We expect the same description to be a good approximation of the potential energy also at finite temperature as long as the domain wall width is small compared to the domain size. Thus we take $m(x)$ to have the following form:

$$m(x) = \begin{cases} +1 & , \quad -L < x < -\frac{w}{2} \\ -1 & , \quad -\frac{w}{2} < x < \frac{w}{2} \\ +1 & , \quad \frac{w}{2} < x < L \end{cases} \tag{A.61}$$

The Fourier coefficients for this case read

$$a_0 = \frac{1}{2L} \int_{-L}^{L} m(x)dx = \frac{L-w}{L} \tag{A.62}$$

$$a_n = \frac{2}{2L} \int_{-L}^{L} m(x) \cos(k_n \cdot x)\, dx = \frac{-4}{n\pi} \sin\left(\frac{n\pi w}{2L}\right) \tag{A.63}$$

We then obtain the magnetostatic energy

$$E_{\mathrm{MS}} = (2\Lambda)^2 d\pi\lambda \left(2\left(1 - \frac{w}{L}\right)^2 + \sum_{n=1}^{\infty} \left(\frac{-4}{n\pi}\right)^2 \sin^2\left(\frac{n\pi w}{2L}\right) \frac{1 - e^{-\frac{n\pi}{L}d}}{\frac{n\pi}{L}d} \right) \tag{A.64}$$

In agreement with eq. 5 in Kooy & Enz [12] if the latter is multiplied by the sample area $(2\Lambda)^2$. We simplify and divide by the sample volume $(2\Lambda)^2 \cdot d$ to get the energy per unit volume a^3, with a being the atomic lattice constant.

$$e_{\mathrm{MS}} = \pi\lambda \left(2\left(1 - \frac{w}{L}\right)^2 + \frac{8L}{\pi^3 d} \sum_{n=1}^{\infty} \frac{1}{n^3}\left(1 - \cos\frac{n\pi}{L}w\right)\left(1 - e^{-\frac{n\pi}{L}d}\right) \right) \tag{A.65}$$

The other energy terms introduced in section 2.2, all per unit volume a^3, are given as follows.

The domain wall energy becomes, with the total domain wall length $l_w = \frac{2\Lambda}{L}2\Lambda$,

$$e_{\mathrm{Wall}} = \frac{\sigma_w}{L} \xrightarrow{T\to 0} \frac{2J}{L}\,. \tag{A.66}$$

The energy in the applied field following eq. (2.16) is

$$e_{\mathrm{Field}} = -\left(1 - \frac{w}{L}\right) h\,. \tag{A.67}$$

For convenience we add a constant term to offset the energy scale by the energy of the homogeneously saturated sample

$$e_{\text{Con}} = -2\pi\lambda + h \,. \tag{A.68}$$

With this we arrive at the total energy density:

$$\begin{aligned}
e_{tot} = &\frac{\sigma_w}{L} - \left(1 - \frac{w}{L}\right)h - 2\pi\lambda + h \\
&+ \pi\lambda \left(2\left(1 - \frac{w}{L}\right)^2 + \frac{8L}{\pi^3 d}\sum_{n=1}^{\infty}\frac{1}{n^3}\left(1 - \cos\frac{n\pi}{L}w\right)\left(1 - e^{-\frac{n\pi}{L}d}\right)\right)
\end{aligned} \tag{A.69}$$

expanding and collecting terms gives

$$e_{tot} = \frac{\sigma_w}{L} + \frac{w}{L}h - 4\pi\lambda\frac{w}{L} + 2\pi\lambda\frac{w^2}{L^2} + \frac{8\lambda L}{\pi^2 d}\sum_{n=1}^{\infty}\frac{1}{n^3}\left(1 - \cos\frac{n\pi}{L}w\right)\left(1 - e^{-\frac{n\pi}{L}d}\right) \,. \tag{A.70}$$

A.2.1 The range $d < L$

This energy is easy to evaluate except for the infinite sum. Let's have a closer look at it:

$$S_n = \sum_{n=1}^{\infty}\frac{1}{n^3}\left(1 - \cos\frac{n\pi}{L}w\right)\left(1 - e^{-\frac{n\pi}{L}d}\right) \tag{A.71}$$

Using $2\cos x = e^{-ix} + e^{ix}$, expanding the parentheses and via the definition of the trilogarithm (polylog 3)

$$Li_3(x) = \sum_{n=1}^{\infty}\frac{x^n}{n^3} \tag{A.72}$$

the series can be transformed into

$$\begin{aligned}
S_n = &Li_3(1) - Li_3\left(e^{-\frac{\pi d}{L}}\right) - \frac{1}{2}Li_3\left(e^{\frac{i\pi w}{L}}\right) - \frac{1}{2}Li_3\left(e^{-\frac{i\pi w}{L}}\right) \\
&+ \frac{1}{2}Li_3\left(e^{-\frac{\pi d}{L} + \frac{i\pi w}{L}}\right) + \frac{1}{2}Li_3\left(e^{-\frac{\pi d}{L} - \frac{i\pi w}{L}}\right) \,.
\end{aligned} \tag{A.73}$$

In the special case of zero magnetic field ($h = 0$) we can set $w = L$ and by using

$$Li_3(z) + Li_3(-z) = \frac{1}{4}Li_3\left(z^2\right) \tag{A.74}$$

and dropping the $-2\pi\lambda$ introduced in (A.68) the total energy per atom of the system reads

$$e = \frac{\sigma_w}{L} + \lambda\frac{16}{\pi^2}\frac{L}{d}\left[\frac{7}{8}\zeta(3) - \frac{1}{2}Li_3\left(e^{-\frac{\pi d}{L}}\right) + \frac{1}{2}Li_3\left(-e^{-\frac{\pi d}{L}}\right)\right] \,. \tag{A.75}$$

where $\zeta(z)$ denotes the Riemann ζ-function. For the general case $w \neq L$ we go back to (A.73) and write the trilogarithm using the following expansion which is valid for $\mu \in \mathbb{C}$ and $|\mu| < 2\pi$.

$$Li_3\left(e^\mu\right) = \frac{\mu^2}{2}\left(\frac{3}{2} - \ln(-\mu)\right) + \zeta(3) + \zeta(2)\mu + \sum_{k=3}^{\infty} \frac{\zeta(3-k)}{k!}\mu^k \qquad (A.76)$$

In our case the condition $|\mu| < 2\pi$ corresponds to $d < 2L$ and $w < 2L$. While in our case $w < 2L$ is true by definition, the condition $d < 2L$ restricts the validity of the calculation to thin films. We abbreviate the sum

$$S_k(\mu) := \sum_{k=3}^{\infty} \frac{\zeta(3-k)}{k!}\mu^k \qquad (A.77)$$

In (A.73) complex arguments of the trilogarithm only appear as complex conjugates. We calculate

$$Li_3\left(e^{-x+iy}\right) + Li_3\left(e^{-x-iy}\right) = \frac{x^2 - y^2}{2}\left(\frac{6}{2} - (\ln(x - iy) + \ln(x + iy))\right)$$
$$+ixy\left(\ln(x - iy) - \ln(x + iy)\right) + 2\zeta(3) - 2\zeta(2)x + S_k(-x + iy) + S_k(-x - iy) \qquad (A.78)$$

with both $x, y \in \mathbb{R}$ and $x > 0$. The complex logarithm $\ln(z)$ is well defined for $\Re(z) > 0$. With

$$\ln(x + iy) = \ln\left(re^{i\varphi}\right) = \ln(r) + i\varphi = \frac{1}{2}\ln\left(x^2 + y^2\right) + i\arctan\frac{y}{x} \qquad (A.79)$$

(A.78) becomes

$$\frac{x^2 - y^2}{2}\left(\frac{6}{2} - \ln\left(x^2 + y^2\right)\right) + xy\arctan\frac{y}{x} + 2\zeta(3) - 2\zeta(2)x$$
$$+ S_k(-x + iy) + S_k(-x - iy). \qquad (A.80)$$

Using this result in (A.73), noting $Li_3(1) = \zeta(3)$ and simplifying considerably we obtain

$$S_n = \frac{1}{2}\frac{\pi^2 d^2}{L^2}\ln\frac{\pi d}{L} - \frac{1}{4}\frac{\pi^2 w^2}{L^2}\ln\left(\frac{w^2}{L^2}\right) - \frac{1}{4}\frac{\pi^2 d^2}{L^2}\ln\left(\frac{d^2}{L^2} + \frac{w^2}{L^2}\right)$$
$$+ \frac{1}{4}\frac{\pi^2 w^2}{L^2}\ln\left(\frac{d^2}{L^2} + \frac{w^2}{L^2}\right) + \frac{\pi^2 dw}{L^2}\arctan\frac{w}{d} \qquad (A.81)$$
$$- S_k\left(-\frac{\pi d}{L}\right) - \frac{1}{2}S_k\left(\frac{i\pi w}{L}\right) - \frac{1}{2}S_k\left(-\frac{i\pi w}{L}\right) + \frac{1}{2}S_k\left(-\frac{\pi d}{L} + \frac{i\pi w}{L}\right) + \frac{1}{2}S_k\left(-\frac{\pi d}{L} - \frac{i\pi w}{L}\right)$$

To make notation easier we drop the S_k-terms for a moment and call the rest $\tilde{S}_n$. By extracting common factors from the logarithms we get

$$\tilde{S}_n = \frac{1}{2}\frac{\pi^2 d^2}{L^2}\ln\frac{\pi d}{L} - \frac{1}{4}\frac{\pi^2 w^2}{L^2}\ln\left(\frac{w^2}{L^2}\right) - \frac{1}{4}\frac{\pi^2 d^2}{L^2}\ln\left(\frac{d^2}{L^2}\right) - \frac{1}{4}\frac{\pi^2 d^2}{L^2}\ln\left(1+\frac{w^2}{d^2}\right)$$
$$+ \frac{1}{4}\frac{\pi^2 w^2}{L^2}\ln\left(\frac{w^2}{L^2}\right) + \frac{1}{4}\frac{\pi^2 w^2}{L^2}\ln\left(\frac{d^2}{w^2}+1\right) + \frac{\pi^2 dw}{L^2}\arctan\frac{w}{d} \tag{A.82}$$

several terms cancel to give

$$\tilde{S}_n = -\frac{1}{4}\frac{\pi^2 d^2}{L^2}\ln\left(1+\frac{w^2}{d^2}\right) + \frac{1}{4}\frac{\pi^2 w^2}{L^2}\ln\left(\frac{d^2}{w^2}+1\right) + \frac{\pi^2 dw}{L^2}\arctan\frac{w}{d} \tag{A.83}$$

again we rewrite the first ln and regroup the terms.

$$\tilde{S}_n = -\frac{1}{2}\frac{\pi^2 d^2}{L^2}\ln\left(\frac{w}{d}\right) + \frac{1}{4}\frac{\pi^2(w^2-d^2)}{L^2}\ln\left(1+\frac{d^2}{w^2}\right) + \frac{\pi^2 dw}{L^2}\arctan\frac{w}{d} \tag{A.84}$$

It is now time to have a look at the S_k-terms:

$$\tilde{S}_k = -S_k\left(-\frac{\pi d}{L}\right) - \frac{1}{2}S_k\left(\frac{i\pi w}{L}\right) - \frac{1}{2}S_k\left(-\frac{i\pi w}{L}\right) + \frac{1}{2}S_k\left(-\frac{\pi d}{L}+\frac{i\pi w}{L}\right) + \frac{1}{2}S_k\left(-\frac{\pi d}{L}-\frac{i\pi w}{L}\right)$$
$$\tag{A.85}$$

$$= \sum_{k=3}^{\infty}\frac{\zeta(3-k)}{k!}\left[-\left(-\frac{\pi d}{L}\right)^k - \frac{1}{2}\left(\frac{i\pi w}{L}\right)^k - \frac{1}{2}\left(-\frac{i\pi w}{L}\right)^k + \frac{1}{2}\left(-\frac{\pi d}{L}+\frac{i\pi w}{L}\right)^k + \frac{1}{2}\left(-\frac{\pi d}{L}-\frac{i\pi w}{L}\right)^k\right]$$
$$\tag{A.86}$$

$$= \sum_{k=3}^{\infty}\frac{\zeta(3-k)}{k!}c_k \tag{A.87}$$

We proceed by calculating the summand for each k.

for $k=3$ we have $\zeta(0)=-\frac{1}{2}$ and $3!=6$. The only term surviving when expanding the terms with power 3 is

$$c_3 = 3\frac{\pi d}{L}\left(\frac{\pi w}{L}\right)^2 \tag{A.88}$$

For all odd $k>3$ the summand vanished since $\zeta(-2n)=0\;\forall n\in\mathbb{N}$. We are therefore left with the computation of the c_k for even k. We set $k=2m$ and get

$$c_k = c_{2m} = -\left(\frac{\pi d}{L}\right)^{2m} - (-1)^m\left(\frac{\pi w}{L}\right)^{2m} + \frac{1}{2}\left(-\frac{\pi d}{L}+\frac{i\pi w}{L}\right)^{2m} + \frac{1}{2}\left(\frac{\pi d}{L}+\frac{i\pi w}{L}\right)^{2m} \tag{A.89}$$

Using the binomial theorem we obtain

$$c_{2m} = -\left(\frac{\pi d}{L}\right)^{2m} - (-1)^m\left(\frac{\pi w}{L}\right)^{2m}$$
$$+ \frac{1}{2}\sum_{n=0}^{2m}\binom{2m}{n}(-1)^n\left(\frac{\pi d}{L}\right)^n\left(\frac{i\pi w}{L}\right)^{2m-n} + \frac{1}{2}\sum_{n=0}^{2m}\binom{2m}{n}\left(\frac{\pi d}{L}\right)^n\left(\frac{i\pi w}{L}\right)^{2m-n} \tag{A.90}$$

$$c_{2m} = -\left(\frac{\pi d}{L}\right)^{2m} - (-1)^m\left(\frac{\pi w}{L}\right)^{2m} + \frac{1}{2}\sum_{n=0}^{2m}\binom{2m}{n}[(-1)^n+1]\left(\frac{\pi d}{L}\right)^n\left(\frac{i\pi w}{L}\right)^{2m-n} \tag{A.91}$$

Obviously all terms with n odd in the binomial sum vanish. We set $n = 2l$ and rewrite

$$c_{2m} = -\left(\tfrac{\pi d}{L}\right)^{2m} - (-1)^m \left(\tfrac{\pi w}{L}\right)^{2m} + \sum_{l=0}^{m} \binom{2m}{2l} \left(\tfrac{\pi d}{L}\right)^{2l} \left(\tfrac{\pi w}{L}\right)^{2(m-l)} (-1)^{(m-l)} \quad \text{(A.92)}$$

$$c_{2m} = -\left(\tfrac{\pi d}{L}\right)^{2m} - (-1)^m \left(\tfrac{\pi w}{L}\right)^{2m}$$
$$+ \left(\tfrac{\pi w}{L}\right)^{2m} (-1)^m + \left(\tfrac{\pi d}{L}\right)^{2m} + \sum_{l=1}^{m-1} \binom{2m}{2l} \left(\tfrac{\pi d}{L}\right)^{2l} \left(\tfrac{\pi w}{L}\right)^{2(m-l)} (-1)^{(m-l)} \quad \text{(A.93)}$$

$$c_{2m} = \sum_{l=1}^{m-1} \binom{2m}{2l} \left(\tfrac{\pi d}{L}\right)^{2l} \left(\tfrac{\pi w}{L}\right)^{2(m-l)} (-1)^{(m-l)} \quad \text{(A.94)}$$

We have therefore obtained the following exact expression for the $\tilde{S}_k$:

$$\tilde{S}_k = -\frac{1}{4}\frac{\pi d}{L}\left(\frac{\pi w}{L}\right)^2 + \sum_{m=2}^{\infty} \frac{\zeta(3-2m)}{(2m)!} \sum_{l=1}^{m-1} (-1)^{(m-l)} \binom{2m}{2l} \left(\tfrac{\pi d}{L}\right)^{2l} \left(\tfrac{\pi w}{L}\right)^{2(m-l)} \quad \text{(A.95)}$$

By collecting the terms from (A.84) and (A.95) we get

$$S_n = -\frac{1}{2}\frac{\pi^2 d^2}{L^2} \ln\left(\frac{w}{d}\right) + \frac{1}{4}\frac{\pi^2 (w^2 - d^2)}{L^2} \ln\left(1 + \frac{d^2}{w^2}\right) + \frac{\pi^2 dw}{L^2} \arctan\frac{w}{d}$$
$$- \frac{1}{4}\frac{\pi d}{L}\left(\frac{\pi w}{L}\right)^2 + \sum_{m=2}^{\infty} \frac{\zeta(3-2m)}{(2m)!} \sum_{l=1}^{m-1} (-1)^{(m-l)} \binom{2m}{2l} \left(\tfrac{\pi d}{L}\right)^{2l} \left(\tfrac{\pi w}{L}\right)^{2(m-l)} \quad \text{(A.96)}$$

recalling the total energy density (A.70) and abbreviating the sum term by S_{ml} we obtain

$$e_{tot} = \frac{\sigma_w}{L} + \frac{w}{L}h - 4\pi\lambda\frac{w}{L} + 2\pi\lambda\frac{w^2}{L^2}$$
$$+ \frac{8\lambda L}{\pi^2 d}\left[-\frac{1}{2}\frac{\pi^2 d^2}{L^2} \ln\left(\frac{w}{d}\right) + \frac{1}{4}\frac{\pi^2 (w^2 - d^2)}{L^2} \ln\left(1 + \frac{d^2}{w^2}\right) + \frac{\pi^2 dw}{L^2} \arctan\frac{w}{d}\right]$$
$$+ \frac{8\lambda L}{\pi^2 d}\left[-\frac{1}{4}\frac{\pi d}{L}\left(\frac{\pi w}{L}\right)^2 + S_{ml}\right] \quad \text{(A.97)}$$

we simplify

$$e_{tot} = \frac{\sigma_w}{L} + \frac{w}{L}h - 4\pi\lambda\frac{w}{L} + 2\pi\lambda\frac{w^2}{L^2}$$
$$+ \frac{4\lambda d}{L}\left[-\ln\left(\frac{w}{d}\right) + \frac{1}{2}\frac{(w^2 - d^2)}{d^2} \ln\left(1 + \frac{d^2}{w^2}\right) + \frac{2w}{d} \arctan\frac{w}{d}\right]$$
$$- 2\pi\lambda\frac{w^2}{L^2} + \frac{8\lambda L}{\pi^2 d}S_{ml} \quad \text{(A.98)}$$

to arrive at the exact energy density for the striped DFIF in a magnetic field for the case $d < L$:

$$e_{tot} = \frac{\sigma_w}{L} + \frac{w}{L}h - 4\pi\lambda\frac{w}{L} + \frac{4\lambda d}{L}\left[-\ln\left(\frac{w}{d}\right) + \frac{1}{2}\left(\frac{w^2}{d^2} - 1\right)\ln\left(1 + \frac{d^2}{w^2}\right) + \frac{2w}{d}\arctan\frac{w}{d}\right]$$
$$+ \frac{8\lambda L}{\pi^2 d}\sum_{m=2}^{\infty}\frac{\zeta(3-2m)}{(2m)!}\sum_{l=1}^{m-1}(-1)^{(m-l)}\binom{2m}{2l}\left(\frac{\pi d}{L}\right)^{2l}\left(\frac{\pi w}{L}\right)^{2(m-l)} \tag{A.99}$$

A.2.2 The limit $d \ll L$

We proceed by expanding the series in (A.99) in powers of $\frac{d}{L}$.

$$\frac{8\lambda L}{\pi^2 d}S_{ml} = \frac{8\lambda L}{\pi^2 d}\sum_{m=2}^{\infty}\frac{\zeta(3-2m)}{(2m)!}\sum_{l=1}^{m-1}(-1)^{(m-l)}\binom{2m}{2l}\left(\frac{\pi d}{L}\right)^{2l}\left(\frac{\pi w}{L}\right)^{2(m-l)} \tag{A.100}$$

The order in $\frac{d}{L}$ is determined by l in the inner sum. We rewrite

$$\frac{8\lambda L}{\pi^2 d}S_{ml} = \frac{8\lambda L}{\pi^2 d}\sum_{m=2}^{\infty}\frac{\zeta(3-2m)}{(2m)!}(-1)^{(m-1)}\binom{2m}{2}\left(\frac{\pi d}{L}\right)^{2}\left(\frac{\pi w}{L}\right)^{2(m-1)}$$
$$+ \frac{8\lambda L}{\pi^2 d}\sum_{m=2}^{\infty}\frac{\zeta(3-2m)}{(2m)!}\sum_{l=2}^{m-1}(-1)^{(m-l)}\binom{2m}{2l}\left(\frac{\pi d}{L}\right)^{2l}\left(\frac{\pi w}{L}\right)^{2(m-l)} \tag{A.101}$$

Now the first (simple) sum together with its prefactor is of order $\frac{d}{L}$ while the second (double) sum is of order $\left(\frac{d}{L}\right)^3$. We continue only with the first sum calling it $S_{d/L}$.

$$S_{d/L} = \frac{8\lambda L}{\pi^2 d}\sum_{m=2}^{\infty}\frac{\zeta(3-2m)}{(2m)!}(-1)^{(m-1)}\frac{(2m)!}{2!(2(m-1))!}\left(\frac{\pi d}{L}\right)^{2}\left(\frac{\pi w}{L}\right)^{2(m-1)} \tag{A.102}$$

$$= \frac{8\lambda L}{\pi^2 d}\sum_{n=1}^{\infty}\frac{\zeta(3-2(n+1))}{2\,(2n)!}(-1)^{n}\left(\frac{\pi d}{L}\right)^{2}\left(\frac{\pi w}{L}\right)^{2n} \tag{A.103}$$

$$= \frac{4\lambda d}{L}\sum_{n=1}^{\infty}\frac{\zeta(1-2n)}{(2n)!}(-1)^{n}\left(\frac{\pi w}{L}\right)^{2n} \tag{A.104}$$

Next we need to recall some properties relating the ζ-function with the Bernoulli-numbers B_n:

$$\zeta(1-2n) = -\frac{B_{2n}}{2n} \tag{A.105}$$

and

$$B_{2n} = (-1)^{n+1}\frac{2\,(2n)!}{(2\pi)^{2n}}\zeta(2n) \tag{A.106}$$

Combining the two, we get

$$\zeta(1 - 2n) = -\frac{1}{2n}(-1)^{n+1}\frac{2\,(2n)!}{(2\pi)^{2n}}\zeta(2n) = (-1)^n\frac{(2n)!}{n\,(2\pi)^{2n}}\,\zeta(2n) \tag{A.107}$$

Using this identity in (A.104) we obtain

$$S_{d/L} = \frac{4\lambda d}{L}\sum_{n=1}^{\infty}(-1)^n\frac{(2n)!}{n\,(2\pi)^{2n}}\,\zeta(2n)\frac{1}{(2n)!}(-1)^n\left(\frac{\pi w}{L}\right)^{2n} \tag{A.108}$$

$$= \frac{4\lambda d}{L}\sum_{n=1}^{\infty}(-1)^n\frac{(2n)!}{n\,(2\pi)^{2n}}\,\zeta(2n)\frac{1}{(2n)!}(-1)^n\left(\frac{\pi w}{L}\right)^{2n} \tag{A.109}$$

$$= \frac{4\lambda d}{L}\sum_{n=1}^{\infty}\frac{\zeta(2n)}{n}\left(\frac{w}{2L}\right)^{2n} \tag{A.110}$$

We use the definition of the ζ-function and get

$$S_{d/L} = \frac{4\lambda d}{L}\sum_{n=1}^{\infty}\sum_{k=1}^{\infty}\frac{1}{k^{2n}}\frac{1}{n}\left(\frac{w}{2L}\right)^{2n} \tag{A.111}$$

We exchange the sums

$$S_{d/L} = \frac{4\lambda d}{L}\sum_{k=1}^{\infty}\sum_{n=1}^{\infty}\frac{1}{n}\left(\frac{w^2}{k^2 4L^2}\right)^n \tag{A.112}$$

the sum over n can be computed explicitly. Using

$$\sum_{n=1}^{\infty}\frac{1}{n}q^n = -\ln(1-q) \tag{A.113}$$

we obtain

$$S_{d/L} = \frac{4\lambda d}{L}\sum_{k=1}^{\infty}-\ln\left(1-\frac{w^2}{k^2 4L^2}\right) \tag{A.114}$$

Again, the sum can be computed explicitly:

$$\sum_{k=1}^{\infty}-\ln\left(1-\frac{q^2}{k^2}\right) = -\ln\left(\frac{1}{\pi q}\sin(\pi q)\right)\,. \tag{A.115}$$

By recalling the other terms of (A.99) we have now for the total energy to first order in $\frac{d}{L}$.

$$e_{d/L} = \frac{\sigma_w}{L} + \frac{w}{L}h - 4\pi\lambda\frac{w}{L} - \frac{4\lambda d}{L}\ln\left(\frac{2L}{\pi w}\sin\left(\frac{\pi w}{2L}\right)\right)$$
$$+ \frac{4\lambda d}{L}\left[-\ln\left(\frac{w}{d}\right) + \frac{1}{2}\left(\frac{w^2}{d^2} - 1\right)\ln\left(1 + \frac{d^2}{w^2}\right) + \frac{2w}{d}\arctan\frac{w}{d}\right] \tag{A.116}$$

This expression is valid for small $\frac{d}{L}$ However, no restriction to w has been applied so far.

A.2.3 The limit $d \ll w$

For small d we also have $\frac{w}{d} \gg 1$. In this regime we can expand the arctan as:

$$\arctan \frac{w}{d} = \frac{\pi}{2} - \frac{d}{w} + \mathcal{O}\left(\left(\frac{d}{w}\right)^3\right). \tag{A.117}$$

Neglecting higher orders we plug this expression into the total energy (A.116)

$$e_{d/l} = \frac{\sigma_w}{L} + \frac{w}{L}h - 4\pi\lambda\frac{w}{L} - \frac{4\lambda d}{L}\ln\left(\frac{2L}{\pi w}\sin\left(\frac{\pi w}{2L}\right)\right) \tag{A.118}$$

$$+ \frac{4\lambda d}{L}\left[-\ln\left(\frac{w}{d}\right) + \frac{1}{2}\frac{(w^2 - d^2)}{d^2}\ln\left(1 + \frac{d^2}{w^2}\right)\right] + \frac{8\lambda w}{L}\left(\frac{\pi}{2} - \frac{d}{w}\right) \tag{A.119}$$

Some terms cancel and we obtain

$$e_{d/l} = \frac{\sigma_w}{L} + \frac{w}{L}h - \frac{4\lambda d}{L}\ln\left(\frac{2L}{\pi w}\sin\left(\frac{\pi w}{2L}\right)\right)$$

$$+ \frac{4\lambda d}{L}\left[-\ln\left(\frac{w}{d}\right) + \frac{1}{2}\left(\frac{w^2}{d^2} - 1\right)\ln\left(1 + \frac{d^2}{w^2}\right) - 2\right] \tag{A.120}$$

For $x \ll 1$ the following approximations hold

$$\ln(1 + x) \approx x \quad , \quad \frac{1}{x} - 1 \approx \frac{1}{x}. \tag{A.121}$$

since for small d, $\frac{d}{w} \ll 1$ we can use these approximations and obtain

$$e_{d/l} = \frac{\sigma_w}{L} + \frac{w}{L}h - \frac{4\lambda d}{L}\left[\ln\left(\frac{2L}{\pi w}\sin\left(\frac{\pi w}{2L}\right)\right) + \ln\left(\frac{w}{d}\right) + \frac{3}{2}\right] \tag{A.122}$$

and further

$$e_{d/l} = \frac{\sigma_w}{L} + \frac{w}{L}h - \frac{4\lambda d}{L}\left[\ln\left(\frac{2L}{\pi d}\sin\left(\frac{\pi w}{2L}\right)\right) + \frac{3}{2}\right] \tag{A.123}$$

If we substitute $w = L - \delta$ this expression is equivalent to the result obtained by Kashuba and Pokrovsky [18] except for the constant $\frac{3}{2}$ coming from the slightly different treatment of the dipolar energy and a factor 2 in the term proportional to h that is due to a typo in their paper.

We continue by calculating the derivatives w.r.t. L and w to minimize the energy. First w:

$$\frac{\partial e_{d/l}}{\partial w} = \frac{h}{L} - \frac{2\pi\lambda d}{L^2}\frac{\cos\left(\frac{\pi w}{2L}\right)}{\sin\left(\frac{\pi w}{2L}\right)} \tag{A.124}$$

equating to 0 and multiplying by L gives

$$h = \frac{2\pi\lambda d}{L}\frac{\cos\left(\frac{\pi w}{2L}\right)}{\sin\left(\frac{\pi w}{2L}\right)} \tag{A.125}$$

note that for $h = 0$ this equation gives immediately $w = L$.

Now we take the derivative w.r.t. L:

$$\frac{\partial e_{d/l}}{\partial L} = -\frac{\sigma_w}{L^2} - \frac{wh}{L^2} + \frac{4\lambda d}{L^2}\left[\ln\left(\frac{2L}{\pi d}\sin\left(\frac{\pi w}{2L}\right)\right) + \frac{1}{2}\right] + \frac{2\pi\lambda dw}{L^3}\frac{\cos\left(\frac{\pi w}{2L}\right)}{\sin\left(\frac{\pi w}{2L}\right)} \tag{A.126}$$

We equate to zero and multiply by L^2:

$$0 = -\sigma_w - wh + 4\lambda d\left[\ln\left(\frac{2L}{\pi d}\sin\left(\frac{\pi w}{2L}\right)\right) + \frac{1}{2}\right] + \frac{2\pi\lambda dw}{L}\frac{\cos\left(\frac{\pi w}{2L}\right)}{\sin\left(\frac{\pi w}{2L}\right)} \tag{A.127}$$

Now we use (A.125) in (A.127):

$$0 = -\sigma_w - wh + 4\lambda d\left[\ln\left(\frac{2L}{\pi d}\sin\left(\frac{\pi w}{2L}\right)\right) + \frac{1}{2}\right] + wh \tag{A.128}$$

Solving for w gives

$$w = \frac{2L}{\pi}\arcsin\left(\frac{L_0}{L}\right) \tag{A.129}$$

with L_0 as given by (2.27). We substitute this result back into (A.125):

$$h = \frac{2\pi\lambda d}{L}\frac{\cos\left(\arcsin(\frac{L_0}{L})\right)}{\frac{L_0}{L}} \tag{A.130}$$

$$\arccos\left(\frac{hL_0}{2\pi\lambda d}\right) = \arcsin\left(\frac{L_0}{L}\right) \tag{A.131}$$

Solving for L gives

$$L = L_0\frac{1}{\sqrt{1 - \frac{h^2}{h_{C,S}^2}}} \tag{A.132}$$

with

$$h_{C,S} = 2\pi\lambda d\frac{1}{L_0} = 4\lambda e^{\frac{1}{2}}e^{\frac{\sigma_w}{4\lambda d}} \tag{A.133}$$

which is again in agreement with Pokrovsky's result except for the factor 2 already mentioned in the context of (A.123). We can check (A.129) and (A.132) for the limiting cases:

i) $h = 0$ gives trivially $L = L_0$ and $w = L_0$ as expected.

ii) For $h \to h_{C,S}$ we have $L \to \infty$. Taking the limit in (A.129) gives

$$w_C = \lim_{L\to\infty}\frac{2L}{\pi}\arcsin\left(\frac{L_0}{L}\right) = \lim_{L\to\infty}\frac{2L}{\pi}\frac{L_0}{L} = \frac{2}{\pi}L_0 \tag{A.134}$$

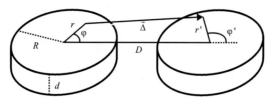

Figure A.1: Geometry for calculating the dipolare interaction between two separated bubbles.

A.2.4 One single stripe domain

From (A.123) we can easily compute the energy of a single stripe in an otherwise homogeneous film. This corresponds to taking the limit $L \to \Lambda \approx \infty$ i.e. $\frac{w}{L} \ll 1$ giving

$$\frac{4\lambda d}{L} \ln\left(\frac{2L}{\pi d} \sin\left(\frac{\pi w}{2L}\right)\right) \approx \frac{4\lambda d}{L} \ln\left(\frac{2L}{\pi d}\left(\frac{\pi w}{2L}\right)\right) = \frac{4\lambda d}{L} \ln\left(\frac{w}{d}\right) . \quad (A.135)$$

The energy per atom then becomes

$$e_{d/l} = \frac{\sigma_w}{\Lambda} + \frac{w}{\Lambda}h - \frac{4\lambda d}{\Lambda}\left[\ln\left(\frac{w}{d}\right) + \frac{3}{2}\right] \quad (A.136)$$

To consider one infinitely long stripe of finite width in an infinite film, the energy per atom is not a meaningful quatity. We multiply by the linear sample size 2Λ to obtain the energy per unit stripe length:

$$E = 2\sigma_w + 2wh - 8\lambda d\left[\ln\left(\frac{w}{d}\right) + \frac{3}{2}\right] \quad (A.137)$$

A.3 Bubble domains - modulation along 2 dimensions

Our aim is to calculate the dipolar energy of a hexagonal lattice of circular bubbles of radius R whose centres have a spacing L under the condition $L > 2R \gg d$ as is expected for the ultrathin limit.

A.3.1 Magnetostatic interaction between two bubbles

We start by calculating the magnetostatic interaction between two flat ($d \ll R$) bubbles B and B' of radius R separated by a distance D, see figure A.1.

The (dipolar) interaction energy is given by (2.19)

$$E_{B\,B'}(D, R) = \frac{\mu_0}{4\pi} \int_0^R \int_0^{2\pi} \int_0^R \int_0^{2\pi} \frac{M\,d\,M'\,d}{|\vec{\Delta}(D, r, r', \varphi, \varphi')|^3} r\,r'\,d\varphi'\,dr'\,d\varphi\,dr \quad (A.138)$$

where Md is the magnetization per unit area and

$$|\vec{\Delta}(D, r, r', \varphi, \varphi')| = |\vec{D} + \vec{r}' - \vec{r}| = \left| \begin{pmatrix} D + r'\cos(\varphi') - r\cos(\varphi) \\ r'\sin(\varphi') - r\sin(\varphi) \end{pmatrix} \right| \tag{A.139}$$

$$= \left[(D + r'\cos(\varphi') - r\cos(\varphi))^2 + (r'\sin(\varphi') - r\sin(\varphi))^2 \right]^{\frac{1}{2}} . \tag{A.140}$$

Note that here we do not have to care about the divergence at $\Delta = 0$ because we only treat the dipolar interaction between the different domains. Therefore $\Delta \geq D - 2R > 0$. The contribution from the interaction within each domain will be taken care of in the next section. We substitute $s = \frac{r}{D}$ and $s' = \frac{r'}{D}$:

$$|\vec{\Delta}(D, s, s', \varphi, \varphi')| = D \left[(1 + s'\cos(\varphi') - s\cos(\varphi))^2 + (s'\sin(\varphi') - s\sin(\varphi))^2 \right]^{\frac{1}{2}} \tag{A.141}$$

$$= D\Delta(s, s', \varphi, \varphi') \tag{A.142}$$

$$E_{BB'}(D, R) = \frac{\mu_0}{4\pi} \int_0^{\frac{R}{D}} \int_0^{2\pi} \int_0^{\frac{R}{D}} \int_0^{2\pi} \frac{D\, M d\, M' d}{[\Delta(s, s', \varphi, \varphi')]^3} s\, s'\, \mathrm{d}\varphi'\, \mathrm{d}s'\, \mathrm{d}\varphi\, \mathrm{d}s \tag{A.143}$$

We expand the integrand as a Taylor-series around $s' = s = 0$:

$$f(s, s', \varphi, \varphi') = \frac{1}{[\Delta(s, s', \varphi, \varphi')]^3} = \sum_{k=0}^{\infty} \frac{1}{k!} \sum_{l=0}^{k} \binom{k}{l} \left. \frac{\partial^{(k)} f}{\partial s^{(k-l)} \partial s'^l} \right|_{s=s'=0} s^{(k-l)}\, s'^l \tag{A.144}$$

As we will see later, the Taylor-coefficients are polynomials in $\cos(\varphi)$, $\cos(\varphi')$, $\sin(\varphi)$ and $\sin(\varphi')$ and are obviously independent of s and s'. We write

$$P_{kl}(\varphi, \varphi') = \left. \frac{\partial^{(k)} f(s, s', \varphi, \varphi')}{\partial s^{(k-l)} \partial s'^l} \right|_{s=s'=0} . \tag{A.145}$$

We now assume that $M = M'$ and recall the definition of $\lambda = \frac{\mu_0}{4\pi} M^2$ (2.20). With this we have

$$\frac{E_{BB'}(D, R)}{D\lambda d^2} = \int_0^{\frac{R}{D}} \int_0^{2\pi} \int_0^{\frac{R}{D}} \int_0^{2\pi} \sum_{k=0}^{\infty} \frac{1}{k!} \sum_{l=0}^{k} \binom{k}{l} P_{kl}(\varphi, \varphi')\, s^{(k-l+1)} s'^{(l+1)}\, \mathrm{d}\varphi'\, \mathrm{d}s'\, \mathrm{d}\varphi\, \mathrm{d}s \tag{A.146}$$

We change the order of integrations and summations getting

$$\frac{E_{BB'}(D, R)}{D\lambda d^2} = \sum_{k=0}^{\infty} \int_0^{2\pi} \int_0^{2\pi} \frac{1}{k!} \sum_{l=0}^{k} \binom{k}{l} P_{kl}(\varphi, \varphi') \int_0^{\frac{R}{D}} \int_0^{\frac{R}{D}} s^{(k-l+1)} s'^{(l+1)}\, \mathrm{d}s'\, \mathrm{d}s\, \mathrm{d}\varphi'\, \mathrm{d}\varphi \tag{A.147}$$

The integrations over s and s' are trivial:

$$\int_0^{\frac{R}{D}} \int_0^{\frac{R}{D}} s^{(k-l+1)} s'^{(l+1)} \, ds' \, ds = \frac{1}{k-l+2} \left(\frac{R}{D}\right)^{(k-l+2)} \frac{1}{l+2} \left(\frac{R}{D}\right)^{(l+2)} \quad (A.148)$$

$$= \frac{1}{(k-l+2)(l+2)} \left(\frac{R}{D}\right)^{(k+4)} = c_{kl} \left(\frac{R}{D}\right)^{(k+4)} \quad (A.149)$$

giving

$$\frac{E_{BB'}(D,R)}{D\lambda d^2} = \sum_{k=0}^{\infty} \left(\frac{R}{D}\right)^{(k+4)} \frac{1}{k!} \sum_{l=0}^{k} c_{kl} \binom{k}{l} \int_0^{2\pi} \int_0^{2\pi} P_{kl}(\varphi, \varphi') \, d\varphi' \, d\varphi \quad (A.150)$$

We proceed now by calculating the P_{kl}'s. Herefore we introduce some abbreviations:

$$A := \Delta(s, s', \varphi, \varphi') \quad (A.151)$$
$$A_\varphi := (1 + s'\cos(\varphi') - s\cos(\varphi))(-\cos(\varphi)) + (s'\sin(\varphi') - s\sin(\varphi))(-\sin(\varphi)) \quad (A.152)$$
$$A_{\varphi'} := (1 + s'\cos(\varphi') - s\cos(\varphi))(\cos(\varphi')) + (s'\sin(\varphi') - s\sin(\varphi))(\sin(\varphi')) \quad (A.153)$$
$$A' := (-\cos(\varphi))\cos(\varphi') + (-\sin(\varphi))\sin(\varphi') \quad (A.154)$$

Using these we can calculate:

$$f = \frac{1}{[\Delta(s, s', \varphi, \varphi')]^3} \qquad\qquad = A^{-3} \quad (A.155)$$

$$\frac{\partial f}{\partial s} = (-3)A^{-4}\frac{\partial A}{\partial s} \qquad = (-3)A^{-4}\frac{A_\varphi}{A} = (-3)A^{-5}A_\varphi \quad (A.156)$$

$$\frac{\partial f}{\partial s'} = (-3)A^{-4}\frac{\partial A}{\partial s'} \qquad = (-3)A^{-4}\frac{A_{\varphi'}}{A} = (-3)A^{-5}A_{\varphi'} \quad (A.157)$$

$$\frac{\partial A_\varphi}{\partial s} = \frac{\partial A_{\varphi'}}{\partial s'} \qquad\qquad\qquad = 1 \quad (A.158)$$

$$\frac{\partial A_\varphi}{\partial s'} = \frac{\partial A_{\varphi'}}{\partial s} \qquad\qquad\qquad = A' \quad (A.159)$$

$$\frac{\partial A'}{\partial s} = \frac{\partial A'}{\partial s'} \qquad\qquad\qquad = 0 \quad (A.160)$$

$$\quad (A.161)$$

Each derivative is therefore a sum of terms T, each one having the form

$$T = A^a \, A_\varphi^b \, A_{\varphi'}^c \, A'^d \, e \qquad (a < 0; \, b, c, d \geq 0; \, a, b, c, d, e \in \mathbb{Z}) \quad (A.162)$$

calculating the derivatives w.r.t. s gives

$$\frac{\partial T}{\partial s} = a \, A^{a-2} \, A_\varphi^{b+1} \, A_{\varphi'}^c \, A'^d \, e + A^a \, b \, A_\varphi^{b-1} \, A_{\varphi'}^c \, A'^d \, e + A^a \, A_\varphi^b \, c \, A_{\varphi'}^{c-1} \, A'^{d+1} \, e \quad (A.163)$$

The derivative is again a sum of terms of the form (A.162). The operation of taking the derivative corresponds therefore to a mapping

$$
(a,b,c,d,e) \xrightarrow{\partial_s}
\begin{cases}
(\quad a-2 \quad, \quad b+1 \quad, \quad c \quad, \quad d \quad, \quad a \cdot e \quad) \\
(\quad a \quad, \quad b-1 \quad, \quad c \quad, \quad d \quad, \quad b \cdot e \quad) \\
(\quad a \quad, \quad b \quad, \quad c-1 \quad, \quad d+1 \quad, \quad c \cdot e \quad)
\end{cases}
\tag{A.164}
$$

and analogously for s'

$$
(a,b,c,d,e) \xrightarrow{\partial_{s'}}
\begin{cases}
(\quad a-2 \quad, \quad b \quad, \quad c+1 \quad, \quad d \quad, \quad a \cdot e \quad) \\
(\quad a \quad, \quad b-1 \quad, \quad c \quad, \quad d+1 \quad, \quad b \cdot e \quad) \\
(\quad a \quad, \quad b \quad, \quad c-1 \quad, \quad d \quad, \quad c \cdot e \quad)
\end{cases}
\tag{A.165}
$$

Evaluating a term T at $s = s' = 0$ as required for the Taylor series yields

$$
A^a\, A_\varphi^b\, A_{\varphi'}^c\, A'^d\, e \big|_{s=s'=0}
$$
$$
= 1 \cdot (-\cos\varphi)^b (\cos\varphi')^c \left[(-\cos(\varphi))\cos(\varphi') + (-\sin(\varphi))\sin(\varphi') \right]^d \cdot e
\tag{A.166}
$$

$$
= e \cdot (-\cos\varphi)^b (\cos\varphi')^c \sum_{f=0}^{d} \binom{d}{f} \left[(-\cos(\varphi))\cos(\varphi') \right]^{d-f} \left[(-\sin(\varphi))\sin(\varphi') \right]^f
\tag{A.167}
$$

$$
= e \cdot \sum_{f=0}^{d} \binom{d}{f} (-\cos\varphi)^{b+d-f} (\cos\varphi')^{c+d-f} (-\sin\varphi)^f (\sin\varphi')^f
\tag{A.168}
$$

Hence, the P_{kl} are polynomials in $(-\cos\varphi)$, $(\cos\varphi')$, $(-\sin\varphi)$ and $(\sin\varphi')$ as anticipated before. To calculate the integrals over the angles φ and φ'

$$
I_{kl} = \int_0^{2\pi} \int_0^{2\pi} P_{kl}(\varphi, \varphi') \, d\varphi' \, d\varphi
\tag{A.169}
$$

we need to know the value of the integral

$$
\int_0^{2\pi} \sin^n x \cos^m x \, dx =
\begin{cases}
\dfrac{(n-1)!!(m-1)!!}{(n+m)!!} \, 2\pi & ,\, n \text{ and } m \text{ even} \\
0 & ,\, \text{if } n \text{ or } m \text{ odd.}
\end{cases}
\tag{A.170}
$$

The result of

$$
\int_0^{2\pi} \int_0^{2\pi} e \cdot \sum_{f=0}^{d} \binom{d}{f} (-\cos\varphi)^{b+d-f} (\cos\varphi')^{c+d-f} (-\sin\varphi)^f (\sin\varphi')^f \, d\varphi' \, d\varphi
\tag{A.171}
$$

is therefore

$$
e \cdot \sum_{f=0}^{d} \binom{d}{f} \frac{(b+d-f-1)!!(f-1)!!}{(b+d)!!} \, 2\pi \, \frac{(c+d-f-1)!!(f-1)!!}{(c+d)!!} \, 2\pi
\tag{A.172}
$$

111

if all of $(b + d - f)$, $(c + d - f)$ and f even and zero otherwise. We write

$$P_{kl} = \sum_{m}^{M_{kl}} A^{a_{mkl}} A_{\varphi}^{b_{mkl}} A_{\varphi'}^{c_{mkl}} A'^{d_{mkl}} e_{mkl} \tag{A.173}$$

Where M_{kl} is the number of terms in a given P_{kl}. Using (A.172) we obtain

$$I_{kl} = 4\pi^2 \sum_{m}^{M_{kl}} e_{mkl} \sum_{f=0}^{d_{mkl}} \binom{d_{mkl}}{f} \frac{(b_{mkl} + d_{mkl} - f - 1)!! \, (c_{mkl} + d_{mkl} - f - 1)!! \, [(f - 1)!!]^2}{(b_{mkl} + d_{mkl})!! \, (c_{mkl} + d_{mkl})!!}$$

$$= \pi^2 i_{kl} \tag{A.174}$$

with this definition of the constants i_{kl} and recalling (A.150) the interaction energy of two bubbles of radius R at a center-to-center distance D becomes

$$E_{B\,B'(D,R)} = D\lambda d^2 \sum_{k=0}^{\infty} \left(\frac{R}{D}\right)^{(k+4)} \frac{1}{k!} \sum_{l=0}^{k} c_{kl} \binom{k}{l} \pi^2 i_{kl} \tag{A.175}$$

$$= \frac{\pi^2 \lambda d^2 R^4}{D^3} \sum_{k=0}^{\infty} \left(\frac{R}{D}\right)^{k} \frac{1}{k!} \sum_{l=0}^{k} c_{kl} \binom{k}{l} i_{kl} \tag{A.176}$$

$$= \frac{\pi^2 \lambda d^2 R^4}{D^3} \sum_{k=0}^{\infty} a_k \left(\frac{R}{D}\right)^{k} \tag{A.177}$$

where we have used

$$a_k = \frac{1}{k!} \sum_{l=0}^{k} c_{kl} \binom{k}{l} i_{kl}. \tag{A.178}$$

Notice that for all k odd a_k vanishes. The numerical values of the a_k are given in table A.1.

The coefficients (a, b, c, d, e) for each P_{kl} can be defined recursively using the mappings (A.164) and (A.165) with the starting point

$$P_{00} = f = A^{-3} \Rightarrow (a, b, c, d, e) = (-3, 0, 0, 0, 1) \tag{A.179}$$

A.3.2 Bubble lattice

The dipolar energy for any fixed bubble (translation invariance in an infinite film!) is composed of the energy of an isolated bubble in a homogeneous background as given by Thiele's formula [22]

$$E_{Dip,BG} = -8\pi R d^2 \lambda \ln\left(\frac{8R}{d\sqrt{e}}\right) \tag{A.180}$$

k	0	2	4	6	8
a_k	1	$\dfrac{9}{4}$	$\dfrac{375}{64}$	$\dfrac{8575}{512}$	$\dfrac{416745}{8192}$
	1	2.25	5.859	16.748	50.8722
$\tilde{S}_k$	$8.3893 \cdot 10^{-1}$	$1.2698 \cdot 10^{-1}$	$3.2540 \cdot 10^{-2}$	$9.4921 \cdot 10^{-3}$	$0.29181 \cdot 10^{-2}$

k	10	12	14	16	18
a_k	$\dfrac{5282739}{32768}$	$\dfrac{552675123}{1048576}$	$\dfrac{29607595875}{16777216}$	$\dfrac{3232491522975}{536870912}$	$\dfrac{44791231022395}{2147483648}$
	161.216	527.072	1764.75	6020.98	20857.5
$\tilde{S}_k$	$9.2134 \cdot 10^{-4}$	$2.9552 \cdot 10^{-4}$	$9.5767 \cdot 10^{-5}$	$3.1259 \cdot 10^{-5}$	$1.0257 \cdot 10^{-5}$

Table A.1: First 10 coefficients in the series expansion for the bubble lattice energy. See text.

plus the interaction with the other domains as derived in the previous subsection.

As depicted in figure A.2 the plane surrounding a specific bubble can be divided into 6 equivalent sectors. The dipolar energy from the interaction with the bubbles in one sector is

$$E_{Dip,sector} = \sum_{n=1}^{\infty} \sum_{m=0}^{\infty} E_{B\,B'}(D_{nm}, R) \tag{A.181}$$

where D_{nm} denotes the distance from the fixed bubble to the bubble at position (n, m) in the sector:

$$D_{nm} = L\left(\left(n^2 + \tfrac{1}{2}m\right)^2 + \left(\tfrac{\sqrt{3}}{2}m\right)^2 \right)^{\frac{1}{2}} = L\left(m^2 + nm + n^2\right)^{\frac{1}{2}} \tag{A.182}$$

We then have for the interaction with one sector

$$E_{Dip,sector} = \sum_{n=1}^{\infty} \sum_{m=0}^{\infty} \frac{\lambda d^2 \pi^2 R^4}{D_{nm}^3} \sum_{k=0}^{\infty} a_{2k} \left(\frac{R}{D_{nm}} \right)^{2k} \tag{A.183}$$

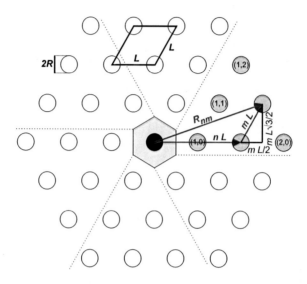

Figure A.2: Representation of the bubble lattice and definition of parameters. The plane around a bubble can be divided into 6 equivalent sectors as indicated by the dotted lines. The bubble lattice constant is L, the bubble radius is R and R_{nm} denotes the position of the bubble (n, m).

Using (A.182) we get

$$E_{Dip,sector} = \sum_{n=1}^{\infty} \sum_{m=0}^{\infty} \frac{\lambda d^2 \pi^2 R^4}{L^3 \left(m^2 + nm + n^2\right)^{\frac{3}{2}}} \sum_{k=0}^{\infty} a_{2k} \left(\frac{R}{L \left(m^2 + nm + n^2\right)^{\frac{1}{2}}}\right)^{2k} \quad (A.184)$$

We change the order of the summations

$$E_{Dip,sector} = \frac{\lambda d^2 \pi^2 R^4}{L^3} \sum_{k=0}^{\infty} a_{2k} \left(\frac{R}{L}\right)^{2k} \sum_{n=1}^{\infty} \sum_{m=0}^{\infty} \frac{1}{\left(m^2 + nm + n^2\right)^{\frac{3+2k}{2}}} \quad (A.185)$$

The double sums over n and m are now independent of any model parameters and and can be computed numerically. Their values S_{2k} are given by

$$S_{2k} = 1 + \sum_{n=1}^{\infty} \sum_{m=1}^{\infty} \left(m^2 + nm + n^2\right)^{-\frac{3+2k}{2}} = 1 + \tilde{S}_{2k} \quad (A.186)$$

The values for $\tilde{S}_{2k} = (S_{2k} - 1)$ are stated in table A.1. The total dipolar energy for one bubble reads thus:

$$E_{Dip,1bubble} = E_{Dip,BG} + 2 \cdot 6 \cdot E_{Dip,sector} \quad (A.187)$$

The factor 6 comes from the number of sectors and the extra factor 2 compensates for the opposite magnetization assumed also at the bubble positions in $E_{Dip,BG}$. To get the energy density, we divide by the volume of the unit cell $V = \frac{\sqrt{3}}{2}L^2 d$.

$$e_{Dip} = \frac{2}{L^2\sqrt{3}d} \left[-8\pi Rd^2\lambda \ln\left(\frac{8R}{d\sqrt{e}}\right) + \frac{12\lambda d^2\pi^2 R^4}{L^3} \sum_{k=0}^{\infty} a_{2k} \left(\frac{R}{L}\right)^{2k} S_{2k} \right] \quad (A.188)$$

This representation of the dipolar energy density is very suitable for a numerical evaluation since the sum over k converges rather fast. This is not obvious because the non-vanishing a_k grow exponentially and for large k we have asymptotically $a_{2k} \rightarrow \alpha^k$ with $\alpha < \frac{7}{2}$ whereas $S_{2k} \rightarrow 1$ for large k. The convergence comes from the fact that $R^2/L^2 \leq \frac{\sqrt{3}}{4\pi} < \frac{1}{7}$ for all values of h and therefore we have $a_{2k}S_{2k}\frac{R^{2k}}{L^{2k}} < 2^{-k}$ resulting in a convergence that is faster than the geometric series with $q = \frac{1}{2}$. This allows us to give an upper bound for the relative error of the result if we consider the sum up to $k = 2K$:

$$1 - \frac{\sum_{k=0}^{2K} a_{2k}S_{2k}\left(\frac{R}{L}\right)^{2k}}{\sum_{k=0}^{\infty} a_{2k}S_{2k}\left(\frac{R}{L}\right)^{2k}} < 1 - \frac{\sum_{k=0}^{K}\left(\frac{1}{2}\right)^{k}}{\sum_{k=0}^{\infty}\left(\frac{1}{2}\right)^{k}} = \left(\frac{1}{2}\right)^{K+1} \quad (A.189)$$

For $K = 10$ we have that the result is exact to at least $5 \cdot 10^{-4}$.

A.3.3 One single bubble

starting from the energy of a single reversed bubble [22],

$$E = 2\pi Rd\left(\sigma_w - 4\lambda d \ln\left(\frac{8R}{d\sqrt{e}}\right) + Rh\right) \quad (A.190)$$

we minimize with respect to R:

$$\frac{\partial E}{\partial R} = 2\pi d\left(\sigma_w - 4\lambda d \ln\left(\frac{8R}{d\sqrt{e}}\right) - 4\lambda d + Rh\right) = 0 \quad (A.191)$$

$$\sigma_w - 4\lambda d \ln\left(\frac{8R}{d\sqrt{e}}\right) - 4\lambda d + Rh = 0 \quad (A.192)$$

i) h_{collapse}

$$\frac{\partial^2 E(R,h)}{\partial R^2} = -8\pi\lambda d^2\frac{1}{R} + 4\pi hd = 0 \quad (A.193)$$

$$\Rightarrow R_{\text{collapse}} = \frac{2\lambda d}{h} \quad (A.194)$$

A Energy calculations

We plug this into (A.192) and equate to zero obtaining

$$\sigma_w - 4\lambda d \ln\left(\frac{8}{d\sqrt{e}}\frac{2\lambda d}{h}\right) - 4\lambda d + \frac{2\lambda d}{h}h = 0 \qquad (A.195)$$

$$\sigma_w - 4\lambda d \ln\left(\frac{16\lambda}{\sqrt{e}h}\right) = 0 \qquad (A.196)$$

$$\ln\left(\frac{16\lambda}{\sqrt{e}h}\right) = \frac{\sigma_w}{4\lambda d} \qquad (A.197)$$

solving for h gives h_{collapse}.

ii) h_c We equate the energy (2.52) to zero:

$$E = 2\pi Rd\left(\sigma_w - 4\lambda d \ln\left(\frac{8R}{d\sqrt{e}}\right) + Rh\right) = 0 \qquad (A.198)$$

since $R \neq 0$ we can simplify

$$\left(\sigma_w - 4\lambda d \ln\left(\frac{8R}{d\sqrt{e}}\right)\right) + Rh = 0 \qquad (A.199)$$

From (A.191) we get

$$2\pi d\left(\sigma_w - 4\lambda d \ln\left(\frac{8R}{d\sqrt{e}}\right) - 4\lambda d + 2Rh\right) = 0 \qquad (A.200)$$

$$\left(\sigma_w - 4\lambda d \ln\left(\frac{8R}{d\sqrt{e}}\right)\right) = 4\lambda d - 2Rh \qquad (A.201)$$

using (A.201) in (A.199) gives

$$4\lambda d - 2Rh + Rh = 0 \qquad (A.202)$$

$$R = \frac{4\lambda d}{h} \qquad (A.203)$$

We use this result to substitute R in (A.199):

$$\sigma_w - 4\lambda d \ln\left(\frac{8}{d\sqrt{e}}\frac{4\lambda d}{h}\right) + \frac{4\lambda d}{h}h = 0 \qquad (A.204)$$

$$\sigma_w - 4\lambda d \ln\left(\frac{32\lambda}{\sqrt{e}h}\right) + 4\lambda d = 0 \qquad (A.205)$$

$$\ln\left(\frac{32\lambda}{\sqrt{e}h}\right) = \frac{\sigma_w}{4\lambda d} + 1 \qquad (A.206)$$

solving for h gives h_c.

Bibliography

[1] A. Aharoni: *Introduction to the Theory of Ferromagnetism.* Oxford University Press, Oxford, 2nd edition (2000)

[2] J. D. Jackson: *Classical Electrodynamics.* John Wiley & Sons, Inc., New York (1999)

[3] W. Heisenberg: Zur Theorie des Ferromagnetismus. *Zeitschrift für Physik* **49**, 619–636 (1928)

[4] C. Kittel: Physical Theory of Ferromagnetic Domains. *Reviews of Modern Physics* **21**, 541–583 (1949)

[5] L. Néel: Anisotropie magnétique superficielle et surstructures d'orientation. *Journal de Physique et Le Radium* **15**, 225–239 (1954)

[6] O. Portmann: *Magnetism in the Ultrathin Limit.* Ph.D. thesis, ETH Zurich (2006)

[7] A. Vindigni, N. Saratz, O. Portmann, D. Pescia, P. Politi: Stripe width and non-local domain walls in the two-dimensional dipolar frustrated Ising ferromagnet. *Physical Review B* **77**, 092414 (2008)

[8] T. Garel, S. Doniach: Phase transitions with spontaneous modulation-the dipolar Ising ferromagnet. *Physical Review B* **26**, 325–329 (1982)

[9] Y. Yafet, E. M. Gyorgy: Ferromagnetic strip domains in an atomic monolayer. *Physical Review B* **38**, 9145–9151 (1988)

[10] R. Czech, J. Villain: Instability of two-dimensional Ising ferromagnets with dipole interactions. *Journal of Physics: Condensed Matter* **1**, 619–627 (1989)

[11] B. Kaplan, G. A. Gehrig: The domain structure in ultrathin magnetic films. *Journal of Magnetism and Magnetic Materials* **128**, 111–116 (1993)

[12] C. Kooy, U. Enz: Experimental and Theoretical Study of the Domain Configuration in Thin Layers of $BaFe_{12}O_{19}$. *Philips Research Reports* **15**, 7–29 (1960)

[13] R. Allenspach, A. Bischof: Magnetization Direction Switching in Fe/Cu(001) Epitaxial Films: Temperature and Thickness Dependence. *Physical Review Letters* **69**, 3385–3388 (1992)

[14] A. Vaterlaus, C. Stamm, U. Maier, M. G. Pini, P. Politi, D. Pescia: Two-Step Dis-ordering of Perpendicularly Magnetized Ultrathin Films. *Physical Review Letters* **84**, 2247–2250 (2000)

[15] O. Portmann, A. Vaterlaus, D. Pescia: An inverse transition of magnetic domain patterns in ultrathin films. *Nature* **422**, 701–704 (2003)

[16] M. Seul, D. Andelman: Domain Shapes and Patterns:The Phenomenology of Modulated Phases. *Science* **267**, 476–483 (1995)

[17] M. Seul, L. R. Monar, L. O'Gorman, R. Wolfe: Morphology and Local Structure in Labyrinthine Stripe Domain Phase. *Science* **254**, 1616–1618 (1991)

[18] A. B. Kashuba, V. L. Pokrovsky: Stripe domain structures in a thin ferromagnetic film. *Physical Review B* **48**, 10335–10344 (1993)

[19] W. F. Druyvesteyn, J. W. F. Dorleijn: Calculations on some periodic magnetic domain structures; consequences for bubble devices. *Philips Research Reports* **26**, 11–28 (1971)

[20] J. A. Cape, G. W. Lehman: Magnetic Domain Structures in Thin Uniaxial Plates with Perpendicular Easy Axis. *Journal of Applied Physics* **42**, 5732–5756 (1971)

[21] K.-O. Ng, D. Vanderbilt: Stability of periodic domain structures in a two-dimensional dipolar model. *Physical Review B* **52**, 2177–2183 (1995)

[22] A. A. Thiele: The Theory of Cylindrical Magnetic Domains. *The Bell System Technical Journal* **48**, 3287–3335 (1969)

[23] A. Abanov, V. Kalatsky, V. L. Pokrovsky, W. M. Saslow: Phase diagram of ultrathin ferromagnetic films with perpendicular anisotropy. *Physical Review B* **51**, 1023–1038 (1995)

[24] O. Portmann, A. Vaterlaus, D. Pescia: Observation of Stripe Mobility in a Dipolar Frustrated Ferromagnet. *Physical Review Letters* **96**, 047212 (2006)

[25] F. H. Stillinger, T. A. Weber: Hidden structure in liquids. *Physical Review A* **25**, 978–989 (1982)

[26] F. H. Stillinger: Relaxation and flow mechanisms in "fragile" glass-forming liq-uids. *The Journal of Chemical Physics* **89**, 6461–6469 (1988)

[27] D. Venus, C. S. Arnold, M. Dunlavy: Domains in perpendicularly magnetized ultrathin films studied using the magnetic susceptibility. *Physical Review B* **60**, 9607–9615 (1999)

[28] J. Schmalian, P. G. Wolynes: Stripe Glasses: Self-Generated Randomness in a Uniformly Frustrated System. *Physical Review Letters* **85**, 836–839 (2000)

[29] C. B. Muratov: Theory of domain patterns in systems with long-range interactions of Coulomb type. *Physical Review E* **66**, 066108 (2002)

[30] H. Vogel: The law of the relation between the viscosity of liquids and the temperature. *Physikalische Zeitschrift* **22**, 645 (1921)

[31] G. S. Fulcher: ANALYSIS OF RECENT MEASUREMENTS OF THE VISCOSITY OF GLASSES. *Journal of the American Ceramic Society* **8**, 339–355 (1925)

[32] G. Tammann, W. Hesse: Die Abhängigkeit der Viscosität von der Temperatur bei unterkühlten Flüssigkeiten. *Zeitschrift fr anorganische und allgemeine Chemie* **156**, 245–257 (1926)

[33] H. Westfahl, J. Schmalian, P. G. Wolynes: Self-generated randomness, defect wandering, and viscous flow in stripe glasses. *Physical Review B* **64**, 174203 (2001)

[34] K. Binder, W. Kob: *GLASSY MATERIALS AND DISORDERED SOLIDS An Introduction to Their Statistical Mechanics.* World Scientific, Singapore (2005)

[35] M. Rubinstein, D. R. Nelson: Order and deterministic chaos in hard-disk arrays. *Physical Review B* **26**, 6254–6275 (1982)

[36] Y. J. Wong, G. V. Chester: Monte Carlo study of glassy order in two-dimensional Lennard-Jones systems. *Physical Review B* **35**, 3506–3523 (1987)

[37] G. Tarjus, S. A. Kivelson, Z. Nussinov, P. Viot: The frustration-based approach of supercooled liquids and the glass transition: a review and critical assessment. *Journal of Physics: Condensed Matter* **17**, R1143–R1182 (2005)

[38] U. Ramsperger, A. Vaterlaus, P. Pfäffli, U. Maier, D. Pescia: Growth of Co on a stepped and on a flat Cu(001) surface. *Physical Review B* **53**, 8001–8006 (1996)

[39] U. Ramsperger: *Structural and Magnetic Investigations of Ultrathin Microstructures.* Ph.D. thesis, ETH Zurich (1997)

[40] A. Vaterlaus, O. Portmann, C. Stamm, D. Pescia: Domain configuration in perpendicularly magnetized atomically thin iron particles. *Journal of Magnetism and Magnetic Materials* **272-276**, 1137–1139 (2004)

[41] K. Koike, K. Hayakawa: Observation of magnetic domains with spin-polarized secondary electrons. *Applied Physics Letters* **45**, 585–586 (1984)

[42] M. R. Scheinfein, J. Unguris, M. H. Kelley, D. T. Pierce, R. J. Celotta: Scanning electron microscopy with polarization analysis (SEMPA). *Review of Scientific Instruments* **61**, 2501–2527 (1990)

[43] R. Allenspach: Spin-polarized scanning electron microscopy. *IBM Journal of Research and Development* **44**, 553–570 (2000)

[44] N. F. Mott: The Scattering of Fast Electrons by Atomic Nuclei. *Proceedings of the Royal Society of London. Series A, Containing Papers of a Mathematical and Physical Character (1905-1934)* **124**, 425–442 (1929)

[45] T. J. Gay, F. B. Dunning: Mott electron polarimetry. *Review of Scientific Instruments* **63**, 1635–1651 (1992)

[46] J. Unguris, D. T. Pierce, A. Galejs, R. J. Celotta: Spin and Energy Analyzed Secondary Electron Emission from a Ferromagnet. *Physical Review Letters* **49**, 72–76 (1982)

[47] J. Dowden, P. Kapadia, G. Brown, H. Rymer: Dynamics of a Geyser Eruption. *Journal of Geophysical Research* **96**, 18059–18071 (1991)

[48] W. Platow, A. N. Anisimov, M. Farle, K. Baberschke: Magnetic Anisotropy and the Temperature Dependent Magnetic Order-Disorder Transition in Fe/Cu(001). *physica status solidi (a)* **173**, 145–151 (1999)

[49] M. Stampanoni, A. Vaterlaus, M. Aeschlimann, F. Meier: Magnetic properties of thin fcc iron films on Cu(001). *Journal of Applied Physics* **64**, 5321–5324 (1988)

[50] J. Giergiel, J. Shen, J. Woltersdorf, A. Kirilyuk, J. Kirschner: Growth and morphology of ultrathin Fe films on Cu(001). *Physical Review B* **52**, 8528–8534 (1995)

[51] H. J. Elmers, J. Hauschild, H. Höche, U. Gradmann, H. Bethge, D. Heuer, U. Köhler: Submonolayer Magnetism of Fe(110) on W(110): Finite Width Scaling of Stripes and Percolation between Islands. *Physical Review Letters* **73**, 898–901 (1994)

[52] D. Z'graggen: *SEMPA Studies of the Stripe Phase in Ultrathin Films of Iron on Copper at Low Temperature.* Master's thesis, Ecole Polytechnique Fédérale de Lausanne and ETH Zurich (2006)

[53] J. Thomassen, F. May, B. Feldmann, M. Wuttig, H. Ibach: Magnetic live surface layers in Fe/Cu(100). *Physical Review Letters* **69**, 3831–3834 (1992)

[54] J. H. Toloza, F. A. Tamarit, S. A. Cannas: Aging in a two-dimensional Ising model with dipolar interactions. *Phys. Rev. B* **58**, R8885–R8888 (1998)

[55] H. Poppa, E. D. Tober, A. K. Schmid: In situ observation of magnetic domain pattern evolution in applied fields by spin-polarized low energy electron microscopy. *Journal of Applied Physics* **91**, 6932–6934 (2002)

[56] A. Bauer, G. Meyer, T. Crecelius, I. Mauch, G. Kaindl: Magnetic stripe domains in ultrathin films and their transformation in magnetic fields. *Journal of Magnetism and Magnetic Materials* **282**, 252–255 (2004)

[57] J. Choi, J. Wu, C. Won, Y. Z. Wu, A. Scholl, A. Doran, T. Owens, Z. Q. Qiu: Magnetic Bubble Domain Phase at the Spin Reorientation Transition of Ultrathin Fe/Ni/Cu(001) Film. *Physical Review Letters* **98**, 207205 (2007)

[58] A. K. Schmid, K. L. Man, N. C. Bartelt, H. Poppa, M. S. Altman: Magnetic Domain Wall Energies in Fe/Cu(100) measured from direct observations of Thermal Fluctuations using SPLEEM. *Microscopy and Microanalysis* **9**, 134–135 (2003)

[59] J. E. Davies, O. Hellwig, E. E. Fullerton, G. Denbeaux, J. B. Kortright, K. Liu: Magnetization reversal of Co/Pt multilayers: Microscopic origin of high-field magnetic irreversibility. *Physical Review B* **70**, 224434 (2004)

[60] D. Vanderbilt: Phase segregation and work-function variations on metal surfaces: spontaneous formation of periodic domain structures. *Surface Science* **268**, L300–L304 (1992)

[61] M. M. Hurley, S. J. Singer: Domain-array melting in the dipolar lattice gas. *Physical Review B* **46**, 5783–5786 (1992)

[62] G. Voronoy: Nouvelles applications des paramètres continus à la théorie des formes quadratiques. *Journal für die Reine und Angewandte Mathematik* **133**, 97–178 (1907)

[63] M. Seul, R. Wolfe: Evolution of disorder in magnetic stripe domains. II. Hairpins and labyrinth patterns versus branches and comb patterns formed by growing minority component. *Physical Review A* **46**, 7534–7547 (1992)

[64] S. A. Langer, R. E. Goldstein, D. P. Jackson: Dynamics of labyrinthine pattern formation in magnetic fluids. *Physical Review A* **46**, 4894–4904 (1992)

[65] W. Helfrich: Conduction-Induced Alignment of Nematic Liquid Crystals: Basic Model and Stability Considerations. *The Journal of Chemical Physics* **51**, 4092–4105 (1969)

[66] J. P. Hurault: Static distortions of a cholesteric planar structure induced by magnetic or ac electric fields. *The Journal of Chemical Physics* **59**, 2068–2075 (1973)

[67] A. Lichtenberger: *Temperature and Applied Magnetic Field Dependence of Magnetic Domains in Ultrathin Fe Films on Cu(001)*. Master's thesis, ETH Zurich (2008)

[68] K. L. Babcock, R. M. Westervelt: Elements of cellular domain patterns in magnetic garnet films. *Physical Review A* **40**, 2022–2037 (1989)

[69] K. L. Babcock, R. M. Westervelt: Avalanches and self-organization in cellular magnetic-domain patterns. *Physical Review Letters* **64**, 2168–2171 (1990)

[70] K. L. Babcock, R. Seshadri, R. M. Westervelt: Coarsening of cellular domain patterns in magnetic garnet films. *Physical Review A* **41**, 1952–1962 (1990)

[71] N. Saratz, T. Michlmayr, O. Portmann, U. Ramsperger, A. Vaterlaus, D. Pescia: Stripe-domain nucleation and creep in ultrathin Fe films on Cu(100) imaged at the micrometre scale. *Journal of Physics D: Applied Physics* **40**, 1268–1272 (2007)

[72] D. Venus, M. J. Dunlavy: Dissipation in perpendicularly magnetized ultrathin films studied using the complex AC susceptibility. *Journal of Magnetism and Magnetic Materials* **260**, 195–205 (2003)

[73] D. Kivelson, S. A. Kivelson, X. Zhao, Z. Nussinov, G. Tarjus: A thermodynamic theory of supercooled liquids. *Physica A* **219**, 27–38 (1995)

Publications

Local-magnetic-field generation with a scanning tunneling microscope
T. Michlmayr, N. Saratz, A. Vaterlaus, D. Pescia, and U. Ramsperger
Journal of Applied Physics **99**, 08N502 (2006)

Also published in

Local-magnetic-field generation with a scanning tunneling microscope
T. Michlmayr, N. Saratz, A. Vaterlaus, D. Pescia, and U. Ramsperger
Virtual Journal of Nanoscale Science & Technology **13**, Issue 17 May 1 (2006)

Stripe-domain nucleation and creep in ultrathin Fe films on Cu(1 0 0) imaged at the micrometre scale
N. Saratz, T. Michlmayr, O. Portmann, U. Ramsperger, A. Vaterlaus and D. Pescia
Journal of Physics D: Applied Physics **40**, 1268-1272 (2007)

Stripe width and nonlocal domain walls in the two-dimensional dipolar frustrated Ising ferromagnet
Alessandro Vindigni, Niculin Saratz, Oliver Portmann, Danilo Pescia and Paolo Politi
Physical Review B **77**, 092414 (2008)

Magnetic field generation with local current injection
Th Michlmayr, N Saratz, U Ramsperger, Y Acremann, Th Bähler, D Pescia
Journal of Physics D: Applied Physics **41**, 055005 (2008)

Temperature-Induced Domain Shrinking in Ising Ferromagnets Frustrated by a Long-Range Interaction
Alessandro Vindigni, Oliver Portmann, Niculin Saratz, Fabio Cinti, Paolo Politi and Danilo Pescia
in *Complex Sciences, Part I*, edited by Jie Zhou. Springer Berlin Heidelberg, pp. 783-786 (2009)

Experimental Phase Diagram of Perpendicularly Magnetized Ultrathin Ferromagnetic Films
N. Saratz, A. Lichtenberger, O. Portmann, U. Ramsperger, A. Vindigni and D. Pescia
Physical Review Letters **104**, 077203 (2010)

Curriculum Vitae

Personal details

name and first names	Saratz, Niculin Andri
born	March 21, 1979 in Chur, Switzerland
citizen of	Pontresina, Switzerland

Education

1986 - 1992	Primary school, Langnau am Albis
1992 - 1999	Kantonsschule Wiedikon, Zürich
Jan 1999	Matura typus B (Latin and English)
1999 - 2000	Work at Warburg Dillon Read, later UBS Warburg
2000 - 2005	Physics student at ETH Zürich
2002 - 2003	Exchange student at Universidad de Granada, Spain
2004 - 2005	Diploma thesis *Magnetic Fields on the Nanometre Scale* at the Laboratory for Solid State Physics of ETH Zürich in the group of Prof. Dr. D. Pescia.
2005 - 2009	Doctoral student and teaching assistant at the Laboratory for Solid State Physics of ETH Zürich in the group of Prof. Dr. D. Pescia.

Dank

An erster Stelle möchte ich meinem Doktorvater Danilo Pescia für sein Vertrauen danken. Wir führten viele angeregte Diskussionen, und auch wenn wir manchmal stritten waren wir uns am Ende meistens einig. Ich bewundere sein theoretisches Verständnis, seine klare Sicht der Dinge und sein Talent, sogar mich zum Rechnen zu bringen. Die grosse Freiheit und Unabhängigkeit, die er mir bei meiner Arbeit gewährt hat, haben viel zu meiner Motivation beigetragen.

Very special thanks go to Prof. Valery Pokrovsky for being a referee on this thesis and for coming the long way from Texas to Zurich to attend the doctoral exam. Auch Prof. Christian Back bin ich sehr dankbar, dass er zugestimmt hat als Korreferent diese Arbeit zu begutachten. Prof. Andreas Vaterlaus und seine Vorlesung über Oberflächenphysik haben mich zum Magnetismus und in die Gruppe von Danilo Pescia gebracht. Sein konstantes Interesse und seine Unterstützung waren von grossem Wert. Es freut mich sehr, dass er sich auch als Korreferent zur Verfügung gestellt hat.

Urs Ramsperger hat mich mit den Experimenten und dem Labor vertraut gemacht. Sein experimentelles Fingerspitzengefühl hat mich immer wieder schwer beeindruckt und ich bin im sehr dankbar für viele gute Tipps zu Japan.

Mit Thomas Michlmayr habe ich während der ersten Hälfte der Diss das Büro geteilt und wir haben auch viele Stunden im Labor zusammen gemessen. Sein grosses Know-how über Elektronik und Klettern war immer sehr wertvoll, im Labor oder beim Bergsteigen in Japan.

Oliver Portmann hat mit seinem Kompendium über Eisen auf Kupfer die Basis gelegt, welche meine Arbeit überhaupt erst möglich gemacht hat. Herzlichen Dank für viele geduldige Erklärungen.

Alessandro Vindigni has taught me a lot about the theory of magnetism and stimulated many experiments with his ideas. Besides that, he led the Tired Skyscrapers to the third place and was available for the occasional beer in the evening.

Thomas Bähler hat die Modifikationen an der Apparatur geplant, gezeichnet und gebaut sowie das ganze Labor in Schuss gehalten. Ausserdem hat er mir die ersten Schritte in LabVIEW beigebracht. Ohne seine Hilfe wäre nie etwas aus dieser Diss geworden.

Urs Maier und dem ganzen Ferrovac-Team gebührt ebenfalls grosser Dank für Bar, Brunni und die hervorragende technische Unterstützung.

Taryl Kirk became my office mate during the second half of my thesis. I wish him good luck with his project.

Während der Diplomarbeit von Andreas Lichtenberger sind viele der hier abgedruckten Messungen entstanden. Vielen Dank für die Gesellschaft im Labor und in

Dresden.

Bei Prof. Mehmet Erbudak bedanke ich mich für manchen heissen Tip, zu Istanbul, dem Labor und dem Leben im Allgemeinen.

Natürlich möchte ich auch allen Freunden und Kollegen am D-PHYS danken, speziell dem AMP-Brett mit den Mitstreitern der ersten Stunde Alex, Andrey, Bruno, Flo und Oliver.

Reto Spöhel hat als WG Kumpel seit der Studienzeit alle Höhen und Tiefen meiner Diss mitverfolgt. Vielen Dank für viele Gespräche über Papers und anderes.

Für den Ausgleich neben Labor und Computer danke ich meinen Freunden, speziell der Klettercrew mit Silena, Harpo, Colibri und Basil sowie Käthi, Lupo und Mistral für die wichtige moralische Unterstützung.

Ganz besonderer Dank gebührt meinen Eltern. Sie haben mich während der ganzen Ausbildung immer unterstützt und die Hoffnung nie aufgegeben, dass auch die Diss irgendwann einmal fertig wird.

Infinitas gracias le debo a mi Lucía, por su cariño, por su paciencia, por su apoyo en todos los respectos y por muchísimas cosas más.